MINERVA.          WM. ROTCH.                    VICTORIA          MARY.

Pl. 3.

N THE ARCTIC OCEAN SEPT. 1871.

# CHILDREN
# OF THE LIGHT

*Books by* EVERETT S. ALLEN

ARCTIC ODYSSEY
*The Biography of Rear Admiral Donald B. Macmillan*

FAMOUS AMERICAN HUMOROUS POETS
*(for young people)*

THIS QUIET PLACE
*A Cape Cod Chronicle*

# CHILDREN OF THE LIGHT

## The Rise and Fall of New Bedford Whaling and the Death of the Arctic Fleet

### EVERETT S. ALLEN

*with illustrations*

Little, Brown and Company—Boston—Toronto

FIRST EDITION

T 10/73

Library of Congress Cataloging in Publication Data

Allen, Everett S
    Children of the light.

    Bibliography: p.
    1.  Whaling--Massachusetts--New Bedford--History.
I.  Title.
SH383.2.A44        338.3'72'950974485        73-9694
ISBN  0-316-03422-3

*Published simultaneously in Canada
by Little, Brown & Company (Canada) Limited*

PRINTED IN THE UNITED STATES OF AMERICA

*To my friend, Aime Tetrault*

# Contents

## V. THE LAST MEETING

# Illustrations

# Maps

# CHILDREN
# OF THE LIGHT

*Concordia* fitting out at G. and M. Howland wharf shortly after her launching in 1867. *Courtesy of the Melville Whaling Room, New Bedford Free Public Library.*

# The Launching

. . . look at the manner in which the people of New England have carried on the whale fishery . . . Neither the perseverance of Holland, nor the activity of France, nor the dexterous and firm sagacity of the English enterprise ever carried this most perilous mode of hard industry to the extent to which it has been pushed by this recent people . . .

—EDMUND BURKE, "Speech on Moving His Resolution for Conciliation with the Colonies."

It is November 7, 1867, at the shipyard of Josiah Holmes and Brother in Mattapoisett, which is not far from New Bedford, Massachusetts. In terms of American history, this is ancient countryside, and pleasant as well. The water is clean; the land fair; the dry leaves rattle in the shoreline scrub oaks, the bay is unbelievably blue, and a number of important people have come to this place of wood chips, plain buildings, and fundamental tools to see a whaleship launched.

The ship is *Concordia*, built for the Quaker brothers George, Jr., and Matthew Howland who, having labored in concord as owner-agents of whaleships for four decades without need for formal articles of partnership, having shared their countinghouse for all these years in harmony and mutuality of interest, caused this vessel to be built and named—as a symbol of faith in the future, of course, but as a testament to their peaceful past as well. *Concordia*, including outfitting, cost $100,000, reportedly the highest such figure in New Bedford's entire whaling history, which, in 1867, was already more than a century old.

Because these were Quaker times in Quaker country, the matter of the shipyard which produced *Concordia* is worth pondering, as it reflected the ethic and the interests of the contemporary society that built her.

The yard stood first of all for perfection of product, on which its reputation and its existence depended; perfection was not only demanded by the customer, but insisted upon by the company of producers, down to the last and least workman. This may not have been so in every walk of life in that era, but when a man has some part in putting together a ship, even a very minor part, the job carries with it a nagging reminder that if his work is sloppy, somebody—very possibly somebody whom he knows—may sail away and not come back.

This basic assumption of responsibility, directly related as it was to long apprenticeships and pride of craft, produced a fierce stratification among those who built and maintained the ships. Even as the Quaker implicitly accepted social stratification as inevitable (the Chosen Few, the unchosen many), he also acknowledged that masters in some crafts were socially above those in others.

Begin then with the matter of the wood required for the construction of a whaleship, which must be thoughtfully selected by someone who knows precisely what the piece will look like when it is fitted, to what and with what it will be fastened, and what will be demanded of it. Many things, some almost in contradiction, are asked of the elements that make up a ship such as *Concordia*. Since she is destined for Arctic operation in season, her parts and their pattern must enable her to be handled quickly, so that sudden danger may be avoided. Yet within, her dimensions must allow for cargo of large bulk and be sufficiently ample to accommodate a good-sized crew of officers and men for a voyage lasting years in all or most of the earth's several climates. The ship must be stiff enough to bear up under a full gale and to allow the hoisting aboard of massive chunks of whale carcass, yet sufficiently limber to avoid serious strain upon her hull when doing so. (The bark *Willis*, Captain Bradford C. Briggs, whaling on the Atlantic grounds in the fall of 1865, was badly strained while cutting in in bad weather, and was condemned at Fayal in the following year.) In sum, a whaleship demands a little of everything that is practical. She is not quite like any other ship.

A great variety of timber for shipbuilding came out of the forests of Mattapoisett and neighboring Rochester, Acushnet, and Dartmouth and, granted a commission to build a vessel, the yard owner himself hiked through woods and swamps to find precisely what he wanted for the project. In some regions today, the instinct and the idiom involved in such a personal search for the one piece of raw material that is better than all others are still appreciated—but not in many. If you can place your hand upon a living tree, observe its shape and height, study its bend and bark, reflect upon its age and the soil beneath it, and visualize the character of its grain and cross-section—if you can do all this, motivated instinctively by a desire to build better this time than you have ever built, then you know how it was.

The matter of selecting timber for framing was a prime consideration. When the shipbuilder went looking for frame wood, that is, for the durable stuff of which the vessel's ribs are made, he sought pasture oak. This is because a lone oak tree, with no protection from wind in any direction, has sapwood evenly distributed around its heart. Sapwood is tougher than heart, and timber made from such an oak is clearer.

When the builder found the wood he wanted, he marked the trees. Gangs of timber cutters followed; the felled trunks were snaked out with oxen when the ground was frozen or covered with snow, and hauled to the shoreside. There, the logs and later, hewn timbers, were sorted for their roles according to shape and kind—ribs, keel, keelson, frames, planking and sheathing; white and yellow oak, white and pitch pine, chestnut and locust; small parts, large parts, fancywork and tree nails.

To keep the pieces readily accessible, the huge logs—curved, knees, and straight, depending on prospective function—were often scattered over as much as an acre in the shipyard. The wood was laid on blockings and yard hands turned it every day or two, sometimes oftener, depending on the weather, to reduce the effect of sun or distribute it evenly.

Once the half-model of the ship-to-be was made (bread and butter style, of thin boards doweled together), the model's lines were projected to full size and laid out on the loft floor. Then the wooden patterns of the principal shapes made from these lines, and the lines from these patterns, were marked on the selected timber. Then hewers and shapers—perhaps more than half a hundred of them— went to work. Their task was primary; since the tools of the day were neither sophisticated nor mechanized, the ultimate shape of the wood, apart from nature's contribution, was the result of a highly skilled, chip by chip operation more dependent on individual judgment than on any elaborate measuring, guiding, or cutting devices.

When the ship's keel was completed—it was generally of two parallel pieces bolted together, roughly three feet thick and a hundred feet or better long—and the stem and sternpost fastened to it, the hewers shaped and put up the closely set timbers, each about a foot square at the keel.

The shipyard worker most esteemed was the dubber, or wielder of the adze, which is a cutting tool with its blade at right angles to the handle, used with a sweeping blow for dressing timber. The adzeman sharpened his keen tool religiously after every eighty strokes; it was he who put the basic shape upon the wood, who translated rough into square. Most men, even good broad-axe hewers, cannot handle an adze adroitly. It takes a good eye, a deft hand, a consistently well-controlled stroke. There is a good deal more for the dubber to watch out for than simply taking off too much wood that cannot be put back.

Once the frames were installed, the three-inch-thick planking was softened in the yard's steam boxes so it could be bent to the ship's curves, and the augermen swarmed the vessel's length, boring holes for the plank fastenings—spikes, bolts, and tree nails (called "trunnels").

The trunnel was a superlative device, an ingeniously contrived wooden nail, usually of white oak or locust. It was square on one end, gradually turning to round at the other; it was driven into the plank far enough so that part of the square portion was embedded and thus would not turn or loosen. The trunnel head was sawed off flush with the plank, split slightly with a chisel, and a wooden wedge driven in. This fastening was more durable than iron and could be removed only by boring it out. Even poets, in due course, took note of it; in "The Shipbuilders," Whittier advised, "Lay rib to rib and beam to beam, and drive the tree nails free." Leave it to the Yankee Quaker to find a use for a square peg in a round hole.

Now came the caulkers, with the twisted and tarred yarn called oakum, which they drove with hammer and caulking iron into the seams between the planks to make the vessel tight. In *Birth of a Whaleship*, Reginald B. Hegarty wrote, "The caulker can tell by the feel of things just about how much oakum to gather for each blow. His finger not only has to gather the right amount and hold it over the seam for driving, his finger also has to give it a peculiar sort of twist called 'coiling' so the fibers will lie partly across the seam, otherwise the iron would split through the strand instead of driving it into the seam. It is this left forefinger and twist of the right amount of oakum that constitute the great secret of the old-time caulkers."

Caulkers and riggers seldom worked on a ship simultaneously because the first operation ordinarily came well ahead of the second. Ships were rigged principally after they were launched, and had to be caulked leakproof before they were put overboard. But if the two crafts did work side by side on occasion, they paid no attention to each other. The caulker, in fact, was a breed by himself, the most solitary master craftsman of them all. He was an implicitly trusted man; nobody checked his work—on which virtually everything and everybody depended. Caulking tools, especially the mallets with which the caulking was driven, were sacred within the craft. Their heads of live oak were often banded with German silver, which was kept brightly polished, and the wood of the mallets was kept oiled; the caulking irons were wiped with oil-soaked oakum, so they would not stick in the seams.

The boss rigger regarded his help as useful if they knew their jobs, but nevertheless socially unfit. He wore a broad, heavy belt; hanging from it in leather beckets were a blunt-pointed sheath knife, honed like a razor; half an ox horn, the big end, with a wooden bottom, half-filled with hard grease, and a sliver-bright marlinspike. Boss riggers talked to shipowners, masters, and some mates. They did so formally and without awe.

While all of this shipyard effort was in process—mallet upon iron, blade upon timber, and steam upon plank—a secondary community of essential artisans held to the same time schedule, worked with the same launching date in mind. As what the vessel was to be became increasingly defined—clean and bold lines rising against the New England sky—joiners, cabinet makers, painters, carvers and polishers flocked aboard in splendid orderliness, each in his precise moment, each with a special and respected contribution, to apply the finishing touches.

A few miles away, along New Bedford's bustling riverfront, the forges glowed hot orange from daybreak to dark at Green's iron foundry and James Durfee's blacksmith shop and in a dozen other manufactories where ringing hammers shaped the metal—hook, sheet, strap, plate, eye and bolt—for *Concordia*'s needs. Roper, whaleboat builder, sailmaker, chandler, sparmaker; suppliers of twine, oars, trousers and treacle; creators of tryworks bricks and mortar, and barreled biscuits; shapers of casks, staves and barrelheads, and the makers of cooper's tools; purveyors of sextants, pumps, long glasses and logbooks, all had a hand in this production and labored in its behalf in lofts, over roaring fires or whirling lathes, in shed, bakery, barn, store, mill, and office, depending.

For this was the nature and the business of New Bedford. This is what the city—its chimneys, drays, ovens, warehouses, wharves, banks, and railways—was about: the production, maintenance, and operation of the whaleship.

So it was that from a crucible more uncommon than the principals knew—into which had been introduced money, men, and materials, but of equal importance, more pride and love than the reticent makers and molders could have given tongue to—came the ship *Concordia*.

The reticence of her creators was a fundamental, not only of the people of that era, but of the time and place. It is essential to understand that whereas this launching was an occasion for pride and joy, it was nevertheless a Quaker launching, and the tone, as well as the garb, was gray. There was a decent restraint about the total procedure that matched the dignity and seriousness of the ship's construction. The launching was a good moment certainly, but, in the minds of those whose $100,000 had made it possible, the idea of throwing hats in the air to mark the moment's arrival was absurd.

The New Bedford *Mercury* reported of this new Howland vessel, "The bark's frame is of pine and oak, principally the former, all timbers carefully selected and cut in the vicinity of Mattapoisett. Her covering outside and inside is of yellow pine, as are also her deck frame and the covering of her lower deck. Her length is 128.5 feet, her beam, 29 feet; depth, 17 feet, and tonnage, about 300.

"If she does not prove at once a comfortable vessel in the roughest sea and one of extraordinary speed, there is no trust to be placed in the skill of builders or the judgment of seamen.

"The cabins are furnished with perfect neatness and in all their arrangements with an eye to the comfort and convenience of the officers and they are paneled and grained in imitation of curled maple, rosewood, and satinwood, by Caleb P. Purrington of Fairhaven.

"Her name is not only blazoned on her stern, but beautifully presented in her figurehead, a finely carved image of the goddess Concordia, who leans forward from the bow, one hand clasping to her side the horn of plenty and the other extended and holding an olive branch. This figure, as well as that of a noble spread eagle, which is to adorn the stern, was carved by Henry J. Purrington of Mattapoisett. Any artist in wood might be well proud of the finely rounded arm and delicately chiseled figure of his goddess."

It is now beyond discovering whether the horn of plenty was simple artistic license exercised by Mr. Purrington or a symptom of wishful thinking on the part of the Howland brothers. Traditionally, the Roman goddess of concord and peace is symbolized with two hands joined and two serpents entwined around a herald's staff.

The *Mercury*'s description really did not do justice to the *Concordia*. She was, in fact, one of the most rakish whalers, one of the finest-lined, best-appointed, and smartest-sailing ever to come out of the Port of New Bedford. It was as if George and Matthew, in an instant both splendid and poetic in which they paid less attention than was characteristic to the matter of assets and liabilities, had said,

"This one will be the best, the swiftest, and the prettiest." *Concordia*'s speed could be explained as no more than businesslike; time is money. But she did not have to be so un-Quakerishly pretty. Yet she was.

The building of *Concordia* has to be regarded from this latter-day perspective as nothing short of a gallant gesture. When she went overboard on that November day in 1867, the tide was at the flood, and so was life for George and Matthew Howland. *Concordia*, smoothly down the ways and water newly lapping beneath her magnificent bows—*Concordia*, soon to be gone a world away in quest of fortune—was a symbol of unshakable confidence, of a deep-seated belief in the inevitability of innumerable magnificent tomorrows.

There were not, however, as many magnificent tomorrows as George and Matthew had counted on, in that moment of the launching. As a matter of fact, *Concordia*'s figurehead cornucopia was as unrealistic economically as it was contrived artistically. Nor is the account that follows a recapitulation of whaling's "golden era." Rather, *Children of the Light* is a record of the industry's dramatic downfall. It is an effort to come to terms with the lives of those who, like George and Matthew Howland, gave shape and direction to the industry—and to discover what and who were destroyed by that inexorable pursuit.

The story told in *Children of the Light* is a matter of two plays being spun out simultaneously, several thousand miles and several civilizations apart, each in its especially adapted theater. From season to season, some members of the cast of the first play appear upon the stage of the second. (The reverse never occurs.) But despite formidable barriers of language, climate, and culture, the two casts and their roles are in fierce embrace, certainly to the life, and even to the death.

More, it is obvious, given the advantage of retrospect, that these two groups of players possessed undeniable and remarkable similarities. Each group shared its daily life according to strict notions of self-determination, yet each was inclined to relate the success or failure of its worldly ventures to the dictates of some higher power. Each accepted the thesis that material wealth was most likely to be accumulated through the concerted efforts of friends and relatives, that skill in business or trade ought reasonably to be rewarded, and that the accumulation of wealth carried with it a special responsibility, particularly to the less fortunate.

But enough of such comparisons. Let us begin at the beginning, long centuries before *Concordia* was built and longer still before she perished in the Arctic ice.

# I

"Which was in the beginning..."

# 1.

# From Point Barrow to Icy Point

So is this great and wide sea, wherein are
things creeping innumerable, both small and
great beasts . . . there is that leviathan, whom
Thou has made to play therein.
—PSALMS 104 25:26

The ragged crescent of land with which we are concerned ex-
tends along a northeast-southwest axis and lies just below the southern limit of unnavigable polar ice in what, in time to come, will
be called the Chukchi Sea, a coastal extension of the Arctic Ocean.
Some day, this crescent will be the northwest corner of Alaska, but
there is not yet anyone about who thinks that everything has to be
called something.

What one must remember about this place first is that it is, more
often than not, locked and lumped with frozen ice chunks weighing
several tons that have been shoved aground and slammed ashore by
the westerly gales that roar for days on end. It is a place of windswept rock and gravel, of deep-drifted snow, of flat land and tree-
lessness, where water halts in winter, having been relentlessly transformed from liquid to solid. Here, life winds down in the hard cold
of each winter, but stops short of hibernation on penalty of freezing
to death; the deepest-sleeping creature stirs uneasily after no more
than a few days of slumber, shivers, and goes about the cruel business of staying alive.

Remoteness in this place is not simply geographical; it is a state
of mind—or will become so when there are minds here to entertain
the thought.

Of course, if one chooses to believe the people who eventually
came to live longest on this forbidding rind of land between what

will be Point Belcher and Wainwright Inlet, and in the general area
from Point Barrow southwesterly to Icy Cape, it was not, in fact,
God who created the Arctic. It was the raven, called *tulugaak*, who
created the world, raised the land from the water (although none
too satisfactorily in some sections along the eastern Chukchi), and
arranged to break the twenty-four hours into night and day—a division, incidentally, much less clearly delineated here than in most
places.

They also believed, did these people, that the moon is male, the
sun, female, and that the world is flat, resting on four wooden pillars. Because of this land's uniqueness in being simultaneously breathtaking, violent, unpredictable, monotonous, stubborn, hostile, beautiful, and delicate, the suggestion of such unique factors in its origins
as ravens and wooden pillars seems appropriate, even if not scientific.

In any event, in some spring so long ago that the year had no date
or any month either and when the order of things was undisturbed
by man, who was not then recognized by himself as being of consequence, a bowhead whale, rolling easily in the blue-gray sea under
sky patterns of light and shadow, sensed the opening of the ocean
ice beyond what was to become the Bering Strait, and he swam
northward in response, for a summer of feeding along the edges of
the north polar cap.

It is reasonable, being a matter of desirable perspective, to note
that man is only six feet long and an elephant seldom more than
twelve, but the bowhead, named for the flexible baleen or whalebone that hangs from the roof of his mouth in a dense fringe of several hundred blade-shaped plates, is fifty to seventy-five feet long
and weighs a ton per foot.

His skin is an inch thick and black, except for a white area covering the chin and the front portion of the lower jaws; his head
amounts to a third of his entire bulk. He is streamlined, in the best
and most natural sense of the word, and, for all his great girth, he
moves quietly and does not affront the world's quiet waters through
which he passes.

The bowhead's little eye is shortsighted. But to compensate for
what he cannot see, his delicate ear is tuned to miles and acres of

underwater sound, and when he is bound for the Arctic, he sings in the twenty-below weather of northern spring—three or four notes, rising to something like middle C at the end and resembling the hooting of an owl. He sings to let his fellows know where he is going, and to solicit their company.

He follows the same general course each year ("Whales have roads to go in," said old Captain Charles Chace of the Wing fleet), north in the spring and south in the fall. Making such a passage, he will travel at the rate of five or six miles an hour, frequently spouting (air and vapor, not water, as some men will assume, in the course of things) through paired blowholes seven to eight times between submersions. When he does spout, the spray arcs out to the left and right in a manner almost capricious for anything so large.

Because he is a cold-water whale, the bowhead is fat, carrying a heavier weight of blubber than do other whales. Sperm-whale blubber is only five to six inches thick; that of the bowhead is often nearer eighteen. The blubber, a layer of fatty insulation beneath the skin, is his overcoat against the temperature of the Arctic water and a bowhead's blubber may contain up to twice as many barrels of oil (three hundred and twenty-seven, on one occasion) as that of a sperm whale.

Cruising to the north, the bowhead follows the drift of plankton on which he feeds, usually at depths of twenty-five to fifty feet, although, if frightened, he can dive to more than a thousand, being ingeniously designed to avoid "the bends" to which men are susceptible when subjected to abrupt pressure change. He dives, and is propelled by his tail, or flukes, which are in the horizontal plane; these are often twenty feet or more from tip to tip, and when flailing, they can be dangerous for anything close at hand. Yet this whale is not much of a fighter and has no teeth or jaws for biting.

His fins are short, about five feet long at most, and two to three feet wide. Without separate fingers, they are all that is left of the forelegs his ancestors had when they roamed the land, rather than the sea. Now, he uses them to balance and steer.

The bowhead does not show much above the surface when he is under way and because he has a hollow in his back, his head and his hump sometimes make him look, at a distance, like two whales. He likes to feel deep water beneath him, because he needs maneuvering room in order to be safe and this need, as well as the plankton drift, determines the course he chooses. His navigation equipment is superlative. (Captain George Fred Tilton once observed a bowhead make a radical change of course thirty-six hours away from a solid barrier of ice and expressed the opinion that the whale sensed its existence that far in advance.)

Instead of teeth, the bowhead's hundreds of bone slabs, ranging from less than a foot in length to more than twelve feet (the heaviest, about seven or eight pounds in weight) and secured to the skull by several inches of muscle, serve as a strainer for catching the shrimplike crustacea, jellyfish, and other minute creatures on which he feeds. Sometimes this food looks like brown seed, each piece about the size of a match head; masses of the stuff appear upon the sea surface and, on occasion, create a greasy slick. In time, men—when there are men—will learn to look for the slick to find the whale.

When the bowhead feeds, he lies with his body in a nearly perpendicular position in the water, swimming slowly and scooping up his souplike meal by opening his lower lip, which is divided in the center, spreading the bone which hangs from the skull, and using the latter like a vertical sieve. This bone, or baleen, is covered with stiff hair—bristles that have something of a kink to them and therefore serve as strainers.

The whale fills his mouth with food, closes his lips, forces the water out through slits between the bone and swallows his catch, thousands of little pieces, at each gulp. During feeding periods, the whale remains under water from eighteen to twenty-two minutes, rising about three times an hour to empty and refill his lungs, and remaining on the surface from one and a half to two minutes.

The bowhead cow is larger than the bull. The newborn calf is five or six feet in length and cream-colored, usually possessing a few whiskers around the nose to remind that, once upon a time, whales had hair and did not need blubber to keep them warm. The calves play, as do most calves, even with things or creatures likely to harm them. The adults also have tricks, although not for fun, but to avoid danger. They can, if pressed, swim in reverse for a short distance, or sink with such rapidity as to leave a good-sized hole in the water.

So it was, in this spring of some time before people, that the bowhead sang, fed, and swam north to the Arctic, compelled by the ponderous machinery of the globe and responding to sound and substance in the manner of his forebears and of their forebears, through centuries beyond counting.

Of the Arctic, then, to which he is bound.

It is a place of bitter certainties, foremost among them being weather that is so fickle as to be cruel and benign in the same hour.

Especially in the first half of the year, a single day may bring heavy snow squalls and temperatures ranging from thirty degrees below zero Fahrenheit to fourteen above. For a couple of months in summer, the sun does not set; for more than two months in winter, it is never seen. The matter of perpetual darkness has been variously explained as being due either to the angle of the sun and the fact that the earth is an oblate spheroid, or to hostile supernatural forces. Poets and mathematicians may differ on this but perhaps there is some truth to both theories. Actually, the effect of the darkness is more important to the inhabitants than its cause, for it stills the world over which it hangs and inhibits natural processes. And later, when there are people in this place, it will prove to depress the spirit.

The highest summer temperatures here rise briefly to the sixties and the winter average is thirty-five below. As the range of these figures suggests, a chief characteristic of the area is its volatility; it surges, at times capriciously and explosively, from sunshine to gale, from deep freeze to thaw, and from barrenness to the briefest beauty.

This is an ancient place, layered with yesterday's seashores. Deep within its unforgiving earth is strewn the culch of what was, in time beyond reconstructing. It is a beleaguered coastal plain, flatter than otherwise, flooded and saturated, pounded by surf and ice, yet a dry countryside for all the wetness, having no more than a half-dozen inches of rain and snow in a year's time.

Here are shore, marsh, and tundra. One finds no timber of any size, but driftwood in quantity—spruce, birch, and poplar, with bark and roots often intact, is flung up by the sea and strews the beach, especially in the season of open water. A handful of rivers run west to the sea, dead in winter and rushing in summer. Below ground, the permanent frost goes deep and the land's underground drainage is forever inhibited. This is why the shallow lakes are numerous and early summer inevitably means a soaked landscape, until the upper layers of earth dry out.

Sometimes the waterlogged upper layers of earth come loose and slide downhill over the frozen ground beneath. This process has produced long, smooth slopes in many places, and if the soil will not hold still long enough to allow plants to take root, its surface is scarred by ridges of black rock and pebbles.

In April, the first snowbirds appear and a few big-chested eider ducks as well, in from the lonely open sea and also northbound, harbingers of massive seasonal migrations to come. The sea ice is still lumpy, thrown up in imprisoning but magnificent heaps and piles; floe bergs on the shore, the architecture of wind and water, are in beautiful unworldly forms—concave, convex, arch, overhang, ellipse, free form, mound, block and mushroom, with an occasional empty needle's eye on end, framing the distant sky.

Even now, in the weather's mellowing, the first lead of water opened by the offshore easterly may be as much as five to ten miles from the land. Here, water is trapped or covered, more often than not. Ashore, all movement of water is suspended in winter, there being neither rainfall nor snow from October to May. Now, in April, the good time of the whale's coming, the whitefish are in the deep holes of the streams that come down to the beach. The whitefish, which may weigh fifteen to twenty pounds, is a trout's cousin, well-formed, but of smaller mouth, plainer color, and weaker jaw; he is temporarily confined, because the rivers freeze solidly on the bars, locking him into their pools of icy water.

The sun, now coming bright enough on the snowfields to blind, begins to soften not only the river ice and the wind-driven drifts, but the total nature of land and sea. In the offshore leads, there is the new sound of the bowhead, breathing like a leaky pipe organ.

When the warmer weather begins to come and there is light

longer and longer each day, life surges like blood through a severed artery, pumping itself extravagantly upon land and water. Through May, June, and July, it is as if everything, plant and creature, understands that the whole time allowed for migration, mating, bursting, budding, reproducing, flowering, nesting, spawning and rooting—the total time for being—is virtually no time at all and that everything has to be accomplished and done with, for better or worse, while earth, sea, and sky are, for an instant of suspension, an open hand instead of the customary fist.

Through May and half of June, the whales pass every day into the far northeast, to some ever-frozen privacy where whales do what they do. There is a dignity about them, even at a distance—about their great, smooth dark bulk and the quiet ease with which they thread through the pack ice, wetly blowing.

The eiders, rafting sea rovers and voracious gulpers of shellfish, first are only a few straggling flocks far out over the ice, but by mid-May, both their patterns of flight and their numbers change markedly. Now they appear by the eager hundreds and thousands, close in to the land, strung-out miles of them stitched for a moment against the light sky, big wings pumping, flying belly-low over the shore floes and sandspits until about the first of June.

Here and there, male king eiders drop exhausted and die on the bleak beaches and in the cold black waters of the marsh, worn out from lack of food and the burden of the long flight that, in the name of producing life, dictates to them its inexorable timetable. But mostly, they make it to the breeding grounds, north still, and even beyond what will be Point Barrow when men come. And in July and August, there will be everywhere the flesh-worn females, naked-breasted from having plucked away the down to line their nests. Hereabouts, between Belcher and Wainwright, a few white-fronted brant, some snow geese and loons, eager to get on with their egg-laying and such, do not go so far north, but quit a little short of the distance and let the crowd go on without them.

When the snow starts melting, brilliant in its decay under the new heat, it goes so rapidly that by about the twentieth of May spots of bare ground, in addition to the windswept patches of winter, begin to show along the shore line. The water runs out on the sea ice and on the ice in the streams, and soon the snow on top of everything is a cool soup of slush and water.

The snow gets softer and the surface water deeper until the middle of June, when the ice at the mouths of the small streams lets go and a flood of fresh water gushes out over the sea ice for several miles. This is the beginning of what is perhaps the closest thing in nature to the overwhelming of a Hercules through persistent application of lesser strengths that tend to wear him down. In a short time, the water on top makes holes through the weak places in the ice and runs off, except in hollows where small streams remain, stubbornly eroding.

Gradually, these streams above the ice and the warm current from the south that flows beneath it eat through the softer areas of the pack, but it is still not until July that the ice inside the inshore "ridge" begins to surrender to the assault of higher temperatures. Then, it breaks up into fairly good-sized cakes, finally is no longer an entity at all, but has been reduced to pieces as dissimilar in pattern as snowflakes, and, at last, is wholly afloat and shifting, after months of immobility.

At this point in the year, the sun shines day and night, and although the temperature lowers at times to about the point of freezing, the uninterrupted heat and daylight are too much for most of both old and new ice, and the melting is incessant.

With the breakup of the ocean ice in the middle of summer, the pack splits into floes that float northward with wind and current, and large herds of walrus appear off the coast. The walrus, which is to say, whale horse, is kin to the seal, a fat and wrinkled, ordinarily gentle giant. He is likely to be ten to twelve feet long, weigh two thousand to three thousand pounds, and possess ivory tusks that may be as long as a couple of feet, four inches thick, and weigh ten pounds apiece.

He bellows when frightened or angry, and swims as deep as two hundred feet to dig for shellfish with his tusks. The female walrus gives birth to only one pup at a time, suckles it for about two years until its tusks are long enough to dig for food in the sea bottom, and although generally even-tempered, will fight to the death for its young.

By the last of June, the tundra is nearly free of snow, and narrow leads of water are opening alongshore. The hardy plants indigenous to this high latitude, shallow-rooted vegetation that can live in a highly acid soil, make a brilliant carpeting of tiny, short-stemmed flowers, lichens, and mosses that flourish in abundance during their brief life. Conspicuous among them are the buttercup and dandelion.

Now is the quick and extravagant hour of the blue-spiked lupine, wild crocus, mountain avens, Arctic poppy, and saxifrage. Even some of the bare rock surfaces support brown, gray, or black crustaceous lichens that swell and become soft when wet; some of the black ones, called "rock tripe," are edible; they look something like tripe, but taste less like it. On the tundra, sedges and grasses clump into tough tussocks and mounds, providing food for the grazing animals inland, especially for the caribou.

There is also a small yellow poppy here in quantity and, because men have not yet come to this place, it is now only a flower. But when the people come, those people who will live here, they will name the poppy *tukalukad jaksun*, these also being their words for a small yellow butterfly that appears in the same season. The butterfly arrives as the poppy fades, and man, being at least half-poet in his native state, will choose to relate the two and believe that the poppy is transformed at the end of its stemmed life, takes wings, and flies away. Thus, this poppy is destined for greater things than simply being stepped on by the broad-hooved reindeer.

The juxtaposition of symbols of what would be distinctly separated seasons anywhere else is remarkable. Here is a bank of snow, soft and rotting in the ever-present sunlight, yet snow, nonetheless. But just beyond, are acres of bright flowers, hugging the earth and patterned up and down the easy slopes. There is, moreover, another measure of this contrast—less obvious, but even more convincing, and a firm reminder that, although winter may seem to surrender here, it never really does.

The air is warm. It encourages a certain expansiveness in bird and beast. The sea is increasingly open, light sparkles upon its deep blue, and suggests that hard ice never overlaid it. But if you were to dig into the ground, you would find solid glacial ice—even on this day of flowers and sunlight—at depths varying from two to seven feet. What is more, the ice remains clear and solid for many more feet than you are likely to want to dig. So we have a gay garden on top of an icebox and whatever (in time, whoever, as well) wants to survive in these parts has to acknowledge that above all else.

Yet at the moment, the icebox is below and this is the time of the bearded seal, of salmonberry, blueberry and rich sphagnum moss carpeting the harshness of the liberated coast; of fat and firm-fleshed trout, bellies flashing in the hard-running streams. In terms of food and family, these are the best hours of the year for the caribou, fox, ptarmigan, gyrfalcon, loon, and the round-headed snowy owl of white feet and yellow eyes who soars by day above the marshes.

Of all of these, the caribou seems most particularly of this place and no other. Just as the land shapes people to its demands—if and when there are people—so this land has shaped the caribou to survive here.

He roves the inhospitable ground in herds, a grayish or brownish beast about the size of a donkey, tough and strong. The usually meager grazing is enough for him; probably, like a donkey, he could eat thistles if he had to. His oversized feet are just right for the snow or boggy ground. Both sexes have antlers (no other female deer do) and you can tell him from any other deer because the brow tine of one side of the antler hangs down over the middle of his face. Some have suggested he uses this tine to dig snow away from his winter food of lichens, but he does not; he uses his feet.

The caribou is a good strong swimmer, something that some hooved animals are not, and that is a major asset in a countryside as wet as this one can be, and in which swimming is sometimes the best means of escaping from something (and, in due time, somebody) that is hungry and would like to eat you. Like the bowhead, like the walrus, the caribou usually produces only a single offspring; the Arctic knows what and how many it can sustain. Whatever animal, by accident or design, chooses to violate this unspoken, unwritten edict dies of an empty gut at the trailside and leaves a warning of picked bare bones to those passing, whether or not they heed.

The business of eating and being eaten in this place is as important as in any and, in fact, more critical than in most. There is a reasonably proportionate relationship between the two groups, active and passive, that must prevail irrevocably if disaster is to be avoided. In the warm season, even more than in winter—when time, place and creature are braced in resigned frugality against cold and hunger—the tight and precious balance of flora and fauna, each hastily consuming and being consumed, is more obvious.

The latitude or leisure for trial, the margin for error found in other areas of the earth, is lacking here. As the various wheels that make up the works of a watch are each entitled to place and role, but without variation of any, so everything here, breathing or flowering, is entitled to one specific piece of the circumference of the life circle. The circle is small enough at best; the piece allotted each no more than enough—and, if the season is whimsical or deviate, not enough. In every bloom, munching jaw, gasping gill, and struggling wing against the sun, there is continuous evidence of short supply, intense interdependence, and the fragility of the balance between the essentials. If weather change prolongs or truncates the quality of an hour and thus moves a school, flock or herd, or kills or spawns a bed of vegetation, something else lives or dies because of it.

Time runs. By the middle of August, you can see the bowheads bound out of the Arctic sea, swimming to the southwest along the edge of the pack, a sign that the northern ice is beginning to thicken. Seasonal changes in the latter half of the year are quick; spring comes slowly, as if it were reluctant, but the onset of winter is swift, its grip representing the more normal state of affairs here, and the twenty-four hours of sun in late July give way rapidly to twelve hours of daylight and, in something like no time at all, to total darkness.

As with the spring, fall here is a time of great variations of weather, especially of fog and gales that go on so long, so tediously and relentlessly long, that they whip everything (and, eventually, everybody) into a kind of submission. It is as if the warm and the cold were locked in gargantuan struggle, which they are, although there never is any doubt which side will be the victor.

From early September on, it is increasingly difficult to catch sight of whatever sun is left; day after day continues stormy and thick. The frequent westerlies howl, the sea inundates bar and lowland, and with unbroken miles of Arctic wind behind it, it thumps, thunders, and smashes upon the coast, each charging ribbon of dark wave ending in a roil of foam and freezing spume. Lagoons are largely closed by the end of the month, ice in the freshwater ponds is already ten inches to a foot thick, and the weather very likely may be no better than this until spring. By sometime in November, ice begins to form in the ocean. This is the beginning of the sea's closing, the freezing of the pack that may remain unbroken, not only until the early summer following, but even later, if there is no offshore wind to help the rising temperature dissipate it.

Sometimes in the fall, the gales drive the main pack of ice in toward the land. But if this does not happen, the sea alongshore freezes over comparatively smoothly, except for the small floes that always drift back and forth. Because the open Arctic Sea is virtually tideless—the mean rise and fall is only about two-tenths of a foot—this shore area ice may remain unbroken until midwinter when the heavy, continuing winds from the west drive in the old pack ice to the three-fathom bar that lies parallel to the coast and about one and a half miles from it. Inside this bar, the ice often forms to a thickness of more than five feet, with four fathoms of water below that. One might consider this a possible winter anchorage for a ship, in time to come, except that, periodically, the bar is not enough to hold off the wind-driven pack coming in from the sea. And when the old ice drives in on the land, it comes with terrible force, and nothing can survive its crushing.

When winter settles in, not much moves unless it has to. Apart from the seals in the air holes and the little tom cod, with big eyes and big mouth, which schools northward in large numbers under the ice in January and February, things are quiet. Even the business of eating and being eaten goes on at a subdued tempo.

So much for this place. Or almost, for while so many things must be said about it that are clearly forbidding and inhospitable, it is no more than fair to mention a couple that are frequently breathtaking.

The first is mirage. In the Arctic, there is certainly nothing to compare to Sicily's *Fata Morgana*, because there are no castles, trees, or men, and no City of Messina, either. Because the air over the Arctic sea is cold, however, and the underside of the air above it is warm and serves as a reflecting mirror, there are days and days in spring when the mirages are magnificent. To be sure, there is nothing to be reflected but a view of the ice, yet this ice of the sea pack—jumbled, tossed, crystal-peaked, an intricacy of fantastic planes, angular heaps, and geometric irregularities—challenges and beguiles the eye to follow the unique horizon it creates, and stimulates the mind to see in its casual design hundreds of half-recognizable things. And because the tide level does not change much between high and low, such mirages sometimes remain for hours.

The second is the aurora borealis, which seems to vary somewhat depending on the area. It is more impressive, or even more eerie, if you will, because it is most prevalent in cold weather, when the sea is closed and the land is locked frozen and the stillness is heavy and unbroken. There is, in this time, no sound upon the night and the colors in the sky seem almost overwhelming enough to make noise upon the empty and otherwise monochromatic landscape.

In some areas, the display generally comes out of a dark bank that forms on the northern horizon; it stretches out in long, shooting streamers that gradually work overhead to the opposite side of the sky, waving back and forth, so close to the earth as to appear no more than barely beyond reach. Slowly, they light up the heavens and the snow-covered earth, usually showing only two colors, violet and faint orange, until finally their waves recede and subside.

In other parts, the illumination occurs on virtually every clear night and begins in the northeast or northwest, arising from the same kind of dark low bank of clouds. The lights are never stationary, not even for a second, nor do they take the form of bows or arches, as they often do in other latitudes. Instead, great curtains of light, flashing with all the prismatic colors, are drawn across the sky, rising and changing, often culminating in a corona at the zenith, and falling like a shower of meteoric fire. As the winter advances, these displays often become more brilliant.

So much for cold beauty, but it is beauty for all that. And there is some kind of irony in the fact that when men do come to settle in this place, they will be afraid of the aurora borealis, will turn their backs upon it, will wave knives at it threateningly, being unable or unwilling to face what beauty there is here, which the aurora borealis so dramatically provides.

# 2.

# The Innuit

It is held by the Religious Society of Friends
that God endows every human being with a
measure of His own Divine Spirit, by which
he has revealed Himself to His children in all
generations; that this Spirit, which although in
man, is not of man, is the manifestation in our
human nature of the Eternal Word 'which was
in the beginning' . . .
—Rules of Discipline and Advices,
Society of Friends

In the beginning, the people of the Arctic came from somewhere
to the west—Siberia, most likely—and it was before recorded time
and in an era of picture-writing and stone culture. Indian neighbors
to the south eventually came to call them "Eskimos," eaters of raw
flesh, but their own name for themselves was "Innuit."

Herbert L. Aldrich, who visited them in the late nineteenth century, concluded, "Informally, they referred to themselves as 'nakoo-
ruk,' which means good, and more formally, as 'Innuit,' meaning
'the people.' This would indicate that they have regarded themselves
as the sole inhabitants of the earth world, or perhaps as the 'chosen
people.' "

When these people came here, some went inland or even migrated
far to the eastward, but most hugged the treeless shores of the Arctic
Ocean, which was a primary source of life—hunting ground for
seal, walrus and whale, provider of driftwood for fire and building,
and of drinking water from the fresh ice of the floes.

These are robust and healthy people, fairer than the Indian of
North America, with brown eyes and straight black hair. Because
of the hardships they are obliged to endure in their struggle to exist,
most of them die before the age of forty. Large families are rare

among them. There are seldom more than three children and the village death rate frequently is higher than that of births.

For all this, the Innuit of both sexes, in the prime of life, have great powers of endurance. When traveling, or if food is scarce, they eat only once a day and often make a journey of twenty-five miles or more in temperatures far below zero without having had a meal. Generally, they eat two meals daily, in the morning and late afternoon and they hold to the thesis that eating too often makes one get hungry too quickly.

They keep no record of events, although they preserve countless legends. Of mounds marking the site of three huts that were part of an ancient village, they say, "Those were of the time when men talked like dogs." They make ingenious and effective wooden eye shades to guard against painful snow blindness in the spring. Some years ago, a pair of these shades was dug up by accident from twenty-six feet below the ground surface; assuredly, this has been their home for ages.

They are a small, light, and lively people; there is a delicate dimension about their hands and feet, and they move with becoming grace when unencumbered by outer clothing. The men are generally beardless until they are twenty to twenty-five and even then, the hair on their faces is patchy and light. Such whiskers as they do grow are always clipped closely in winter. (Beards collect moisture from the breath and freeze in the hard weather.) For the same reason, the men cut off their eyebrows in the fall and crop their hair as well, although leaving a bang in front.

As the Innuit's life is cruel, he himself is kind; as his weather is violent, he himself is peaceful.

Lieutenant P. H. Ray of the United States Army, who lived here two years with the Innuit, wrote: "They are kind and gentle in disposition and extremely hospitable to strangers. Though they may rob a stranger of every means of obtaining a subsistence in one moment, they will divide with him their last piece of meat in the next.

"They have no form of government, but live in a condition of anarchy; they make no combinations, either for offensive or defensive purposes, have no common enemies to guard against, nor have they any punishment for crimes. I never knew one to attempt to reclaim stolen property though they might see it in the hands of the thief or left on his cache; though given to petty pilfering, they rarely, if ever, break into a cache except into one of meat when driven to it by hunger . . . They never made the slightest resistance to our reclaiming property when discovered and would laugh about it as though it were a good joke.

"They are very social in their habits and kind to each other; we never witnessed a quarrel between men; neither did we ever see a child struck or punished, and a more obedient or better lot of children cannot be found in all Christendom. I never saw anyone of any age do a vicious or mean act."

First Lieutenant D. H. Jarvis, who commanded here an overland expedition of the United States revenue cutter *Bear*, observed, "The hospitality of these people I have never seen equaled anywhere. It is never grudging; it is thrust upon you. The best they have and the best place in the house are at your disposal. It is so universal that it comes as a matter of course and as a result does not seem to be properly recognized or appreciated. Often, it is embarrassing, for the natives are so insistent and generous that it is hard to refuse to accept their offers, and go about your business in your own way.

"Never, in all our journey, did we pass a house where the people did not extend a cordial welcome and urge us to go in, and hardly a hut that we did go into but the best place was cleared out for us and our belongings. What this means to a tired, cold and hungry traveler cannot be fully appreciated save by those who have experienced it and my former good impressions of the Alaskan Eskimo were but intensified by this winter's journey. All that we ever did in return for such hospitality and all that was expected was a cup of tea and a cracker to the inmates of the house after we had finished our meal."

Have they love, then, these people of the Arctic? Not of the kind that relates either to God or romantic poetry. Concerning the first, the Innuit believes in good and bad spirits; the bad one is Toonook, or sometimes, Tuna, and the good spirit is Kelligabuk, which means mastodon. Toonook, who is so powerful as to be both creator and destroyer, who is related to both life and death, is simultaneously in the sea, earth, and air, and he is as well in every suspicious place and sound, even in the wind when it howls.

If you have to go out on a dark night—and it is better not to—if you must journey farther to the south than you have ever been, until all your landmarks are gone behind you—and it is better not to—at least carry your bone knife and have it ready to keep off Toonook. And the same for Kiolya, the aurora borealis; it is a manifestation of evil, but the knife will keep it away. Principally, you have no control over either of these forces, although certainly you can go out of your way to offend them if you are so unwise as to do so—and if you do, not much will go right thereafter unless you get the help of the old and wise in the village, or perhaps of the shaman.

Sometimes, when the whales are late in coming, when the east wind, *nigyu,* does not blow on the ice to open it; sometimes when the eider ducks fly so high or so far out over the pack that they cannot be killed, it is necessary for the old men to drive out Toonook— who is doing these things—by beating drums and speaking in loud voices.

If the shaman must be called, he who derives his magical power from the creature that is his guardian or the creature that he can become at will, he brings his strong symbols of squirrel, wolf, or ptarmigan, and his powerful charms, the raven's head, dried weasel, or loon skin. When the house is darkened, he will beat his drum and sing strong songs. If the soul is stolen, lost, or wandering—and this can bring one misfortune after another—the shaman can find it and restore it to its troubled owner. If there are hostile powers that must be driven away, he will wrestle with them and such is his magical strength that he will defeat them, although ordinary men could not do this.

These then are the gods of the Innuit and these are their relationships to them, and love is not included, although logic is. For they hold the general belief that if you get into trouble, it is probably because you have violated a natural rule. For example, if the whale comes, it is because you have invited him properly, made him want to come, and made him feel welcome. And if you succeed in killing him, it is because he wanted you to, was pleased to allow you to do

Two Eskimo women with chin tattoos. (These and the following three sketches of Eskimo life were executed by a fourteen-year-old Eskimo girl from Point Barrow, probably in the 1890s.) *Courtesy of the Whaling Museum, New Bedford, Mass.*

so because you were nice to him. Similarly, if he does not come, it is because you have offended him, and if you do not kill him, it is because he does not wish you to, for the same reason.

As for mating, love does not figure in that, either.

Sometime when the child is between twelve and fifteen years, childhood ends. In the case of a boy, a stone knife shaped like a chisel is driven through the flesh below the corner of the mouth and a bone labret is buttoned into the cut, so that everyone will know he is no longer an adolescent. In the case of a girl, she is tattooed with parallel lines extending from the center of the lower lip to the base of the chin.

Boys and girls have sexual intercourse before they are married, as biology demands. But promiscuity is discouraged because it is a waste of time; the promiscuous girl is flitting about when she ought to be learning how to keep house. In similar practical vein (although those men or women malformed or of irregular features are less sought after), beauty is held far less important than ability. Obviously, in a land so limited in resources that its people are obliged to devote all their time and energy to the struggle for existence—in a place where they are faced with the constant threat of famine—a male who is a good provider and a female who is a hard worker are far more likely to survive than a pair of handsome incompetents.

There is no marriage ceremony. Sometimes children are betrothed by their parents at an early age; if there has been no childhood engagement, the mother may choose a prospective wife for her son. If she does so, this arrangement also has a practical prelude. The mother invites the girl to her home, where the latter does the housework and the cooking. It is a training and testing period, in which the mother-in-law to be has an opportunity to explain how she likes things done, and in which she can observe whether the young woman is likely to be a good homemaker. During this period, the girl usually returns to her family's home each night until it is time for her to live with her husband.

Although polygamy is infrequent and not generally approved of, and although most husbands and wives remain together, expressions of love as practiced by other races are unknown. As Robert F. Spencer, who made an extensive study of the Alaskan Eskimo, found, "the husband and wife did not talk about anything. There was no intellectual level to marriage. The culture made no provision for expressing feelings, and there was no attempt to evaluate situations or to pass judgments on them. People talked of weather, hunting, and food; no one could feel free to indicate to others that he might be out of sorts. This was true in general relationships and in husband-wife relationships."

Shortly before the woman is to be delivered of a child, she is put in a place apart, a separate house or hut, and it is up to her alone whether either she or the child lives. Old women of the village, the only ones who may come near, place food within her reach and when she has given birth, if the child is alive, the old women take it, roll it in mud or snow, and leave it there naked for a hour or so. In this way, it early becomes inured to the rigors of the climate in which it must live. Or it dies, which is the way of the Arctic.

And when the old women return to take up the child, they howl and chant to drive away the evil spirits that may have been born with it or that hover over and threaten it. Then the young one is sewed up in a skin bag, a blouse and trousers affair, with only its head sticking out. When carried by its mother, the child is sometimes stowed into her gown and, if the atmosphere is more benign, held so that it faces over its mother's shoulder and not back to back. If the baby cries, the mother or some old woman in attendance takes it outdoors and lets it face the breeze. The breeze being what it usually is here, that is enough to stop anybody from crying.

But the time of birth is the only cruel time for the young and its hard aspects are unquestionably related to the unspoken need to eliminate the weak in a country that periodically kills even the strong. Forever after birth, the child is better treated than are children in most of the world; it is suckled at the mother's breast sometimes until it is five or six—and if a favorite offspring, perhaps until ten.

When the young one can sit up, its mother gets on the floor, braces the baby between her outstretched feet and teaches it to swing its arms about, at the sides, and over the head. Then there follow lessons in chanting and dancing, which are important for an adult to know. The former is both simple and complex; the Eskimo language is probably a far-removed offshoot of the Ural-Altaic tongue and as such, each word in its ten-thousand-word vocabulary has its own exact meaning, but the word forms are inflected to fit their use in sentences. The speech has a deep guttural sound with an irregular rhythm and much of the chanting consists of only two notes: *yung ah yah, yung ah*, these sounds being simultaneously precise and flexible enough—depending on intonation—to express both joy and sorrow.

As for dancing, it is of vital importance, being entertainment certainly, but in the larger sense, a pillar of perennial ritual. "Immediately after the departure of the sun," Lieutenant Ray noted, "when food is plentiful, it is customary for each village to hold a three-day carnival. Friends are invited from neighboring villages and the time is passed in dancing, singing and feasting. The *kudyigin*, or council house, is fitted up with a new roof of ice and crowded day and night, fresh dancers taking the places of those tired out, and the dull tum tum of the drum, mingled with snatches of song and shouts of laughter, can be heard coming from every iglu.

"Sometimes, they have performers who do a pantomime with five men and two women, attired in new deerskin suits with the flesh side out and men wearing tall, conical hats of sealskin, ornamented with dentalium shells and tufts of ermine and Arctic fox fur. The women are bare-headed, with their hair neatly plaited. They often have a drummer and two singers, who make a doleful chant, the dancers keep time with their feet, swaying their bodies to left and right with spasmodic jerks. The men, one at a time, spring to the front, and in wild gestures portray how they have taken seal, bear or deer, being cheered by the crowd as they finish and take their place in line.

"Two large stone lamps light up the crowded sixteen-by-twenty room; the people are unwashed, there is no ventilation and the temperature gets up to a stenched eighteen degrees."

The deerskin suits are an institution, born of generations of experience. Winter clothes, all of light and warm deerskin, consist of

Eskimos dancing, winter (top) and summer (bottom). *Courtesy of The Whaling Museum, New Bedford, Mass.*

a pair of tightly fitting trousers, with the hair next to the skin, to damp the cold (and the reverse for more ordinary weather); a pair of socks, with the hair next to the feet; a pair of boots, with the hair outside, and with the soles of heavy sealskin for hard wear and deer-skin for light wear; two *artigges*, or shirts, one with the hair next to the body and the other with the hair outside, and both with close-fitting hoods fringed with wolf skin, to break the wind from the face and nose. Add to this, a pair of mittens. All of these garments are made of the summer skins of the caribou and the whole outfit will not weigh more than twelve pounds.

If this outfit is well sewn and tight, it will defy almost any degree of cold; no amount of woolen clothing would produce the same re-sult. The heavy winter skins of the reindeer are rarely used for clothes, but make good sleeping gear, because they are light and warm; all of the skins are beautifully soft and pliable, being tanned in human urine.

These people have a unique philosophy of sharing and accumu-lating.

Here is a fellow who has been out on the ice this winter morning, hunched into half a gale for hours watching an air hole. Finally, he kills a fat seal. Having dragged the heavy carcass more than five miles to the land, he comes to his house to butcher it. The word is out, because if you are hungry—and hunger is something these people live with, as they live with the ice and it is as much a part of many days as is the westerly blast—you know very quickly who has food.

There are five neighbors waiting for him and a couple more com-ing, and they all want some of the meat. The behavior patterns of this place allow them to select the chunk they want; often enough, they pick the best. The hunter must give to anyone who asks, for one of the worst things that can happen to him is to get a reputation for being stingy. Those who beg are not supposed to take more than they need; asking for enough extra to store in the cache is not con-sidered good manners, but, on the other hand, it is very hard to know whether anybody does this, and even if he is discovered doing so, the successful hunter must continue to give to him and may not even criticize him.

Especially in times of food shortage, it is often the families of the best hunters who go hungry, for once the catch is butchered and in the ice cellar, it belongs to the wife, but until it gets there, everybody else has a right to a piece of it. It is the same way with personal possessions; ownership of anything carries with it a virtual obligation to loan it to anyone upon request. In hard times, the burden upon the wealthier members of the community is heavy; they may very likely have to support and sustain the whole village, thus wiping out whatever food surplus their skill or prudence had created.

So it is very hard to accumulate surplus goods or food, and it is a large responsibility if you do. That is why there are not many wealthy families in this place, and why many men actually avoid becoming wealthy. Yet some do, and it is important to understand how this comes about.

If you are not a good hunter, you are not likely to become wealthy because that is where the whole business starts. Men who acquire no great skill with the spear or harpoon live precariously; some of them may turn to craftsmanship, making tools, weapons, or boats or sledges in exchange for food or something else they need. But this is not a good life and they are likely to find it hard to get wives. Even if you are a good hunter, you also have to have a family if you are going to get ahead; it is significant that the word used here for "poor" means having no family, no kin to whom to turn. Occasionally in the winter, there is held a dance in the council house so that those who have more than enough may give clothes, weapons and tools to the poor. This is not charity because the Eskimo does not have such a word in his language nor is there any stigma attached to receiving such gifts, even if the poverty arises from somebody's laziness. Mostly, these are not lazy people anyway, industry being as highly regarded as generosity.

Last season then, you did well at winter sealing and were the first to strike one of two bowheads taken in the spring, creating a very favorable impression in your family. In addition, since you are an only son, the death of your father recently provided you with all of his hunting gear. In short, you look like a good investment. So all of your family, and perhaps even some non-kin, ask you whether you would like to try to become an *umealiq*, or whaling leader; if so, they will back you by contributing sufficient goods to get you started.

The decision requires considerable thought. Many men would not take the step even if they had the backing because of the demands made upon a whaling captain and the great possibility of failure, with attendant stigma. Still, it is tempting; the *umealiq* is the nearest thing to an aristocrat there is in this society.

You decide to go ahead.

Getting your boat crew is more complicated than taking a wife; it may be an equally long-lasting commitment, and involve about as many responsibilities. Obviously, you want men who are skillful hunters; so does every other whaling captain. Thus, some of your crew—excluding the occasional youth of promise who has not been spoken for previously—must be lured away from *umealiqs* who want them as much as you do. Kinship may be helpful to you in some cases, but basically, it is a matter of bargaining and bribery; a good man will hold out for the best deal and go where he is offered most, as in most societies.

Once acquired, the crew is as much a social unit as it is an economic one. The captain is responsible for the support of his men and their families; since he is wealthier and more proficient than they, he is obliged to be generous, as well as modest, or he will find himself with no crew. Through shrewdness in trading, skill in hunting, leadership that is certain, but unassuming, he has to work constantly to earn and retain their allegiance. This boat team may very likely fraternize and work and hunt together from one season to another; the partnership is closely knit for the obvious reason that the lives and well-being of all those within it hang upon its effectiveness. Essentially, what is asked of the captain is continuing success—he must forever increase his supply of goods and food and one bad season can wipe him out, no matter how skillful he may be.

Assuming that you have made an economic success as a whaling captain, a particular behavior pattern goes with your role. If you do not follow it closely, whatever else you may have achieved will elicit little respect.

Spencer noted that the man of wealth was expected to take advantage of his position so as to increase his holdings but that he was obliged to give them away when asked and to avoid any appearance of niggardliness or envy toward others. "The man of wealth," he wrote, "never boasted of his success; he was modest, generous, and expected to show dignity and serenity. He was not obliged to conceal his wealth, but at the same time, he could not make undue display of it . . .

"The great umealiq, Taakpuq, who lived at Utkeaayvik . . . offers an example of the behavior of a man of wealth. He wore parkas of beautifully matched skins, his labrets were the most expensive and he wore a headband of green and white beads.

"The many whales he had taken were indicated by the tattoo marks on his cheeks.

"He is said to have walked slowly and with great dignity, speaking seldom and then in a measured way and he was noted for his extreme generosity. He never boasted of his successes and he sat silently (in the council house) while others played games and made sport. His was an ideal character, recalled by all who knew or had seen him. It is said that he could enlist the services of any man in the community."

Paradoxically, although success brings its rewards in this society, either too much success or too much effort to be successful is discouraged, as are other forms of aggressiveness. It is good to trade effectively, but if you drive too hard a bargain, nobody will deal with you thereafter. It is fine to hunt well, but not too well, because the shaman dislikes the man who always wins and he will make him ill, or perhaps steal his soul. Men who are wealthy have to be far more careful of offending the shaman in this manner than do people who have less.

Moreover, you may not offer advice unless it is requested. If you are going to boast, you must do so by being self-deprecating, and it is better to lose an argument than to win one, because these people will walk off and leave a quarrelsome person in the middle of a sentence, rather than hear him out. Although they never heard of the expression, they "turn the other cheek" because to do so gains them both approval and prestige. If you refuse to fight with your enemy, you will have won in the eyes of the community. If a man steals your wife and you do not seek revenge—better still, if she returns and you take her back without scolding—your neighbors will look up to you.

This attitude in the direction of peace and reason reflects both the peaceful nature of the Eskimo and his pragmatic attitude toward society. For if a disagreement results in murder, the families of the principals are automatically committed to a feud, and relatives will do everything they can to keep quarrels below the boiling point. In addition, feuds mean an erosion of the community cooperation that is so essential for survival; any great number of major disagreements would fracture the whole system of boat crews and communal hunting and fishing that characterizes most of the year's activities except winter sealing, which is done individually.

The council house is the only public building in an Innuit community. It is owned by the boat captains because they supplied the materials. It is about fourteen by thirty-six feet, made of blocks of ice on a foundation of sod. A short passageway of ice blocks leads into one side; this storm porch keeps out the worst of the weather. The windows are of clear ice and the roof is made of skins, held in place with blocks of ice, and stretched over driftwood beams and strips of whalebone. The sides of the building are banked with snow to keep out the cold. The skylight is formed by sections of translucent whale gut sewed into the skins of the roof cover.

The name of the council house is the name of the community; in this building the men, and sometimes the women, work, play games, conduct their ceremonials, and dance. There are charms on the walls and hanging from the roof beams—a stuffed gull, a whale figure of bone, little men of wood in the skin boats called *umiaks,* in which they go whaling—and when the women bring food to the men here on winter days, the trencher is set before each charm before the meal is given to them. Women may not come in here when they are menstruating or when the boat crews are preparing for whaling.

Hanging from the roof is a ring of hide. This is one of the games they play. The idea is to jump off the floor, hook one finger into the ring, and see how long you can hang by it. Through such games, many of them calling for endurance or coordination, the young men who are invited to the council house attract the kind of attention that enables them to become members of a whaling crew.

In the council house, the young men also join their elders in smoking. The use of tobacco among these people is more than simple pastime or even deeply rooted habit. It is an ancient comfort in a comfortless world; its use is half-ritual; it is sometimes an avenue of flight from the harsh realities, and sometimes an aid to decision-making. Long ago, before they had tobacco, they smoked *kilikinick* (which American whalemen to come, including Captain George Fred Tilton, called "kilicanuck," applying the term to any tobacco that, in their view, was strong enough to kill a French Canadian)— a mixture of the bark of the Arctic willow mixed with catkins. But in later times, there has been a limited quantity of black-leaf Russian tobacco, brought here by way of the Bering Strait and the Diomede Islands. Men, women and children smoke or chew; if they are chewing tobacco while they are also chewing meat to make pemmican of (and they do, more often than not), the pemmican becomes flavored with tobacco juice, which they do not mind at all.

If they smoke, their pipes are generally of stone or walrus ivory, with somewhat cumbersome stems and little shallow bowls. Pipes made of wood are less common, because of the scarcity of wood. It is customary to fill the bottom of the bowl with a pinch of hair from the reindeer, called *tuk-tu,* which when dry is highly inflammable; tobacco about the size of a pea is placed on top of the hair. Fire to light the pipe usually is made with a spruce-shaft drill about eighteen inches long, turned rapidly back and forth with a small bow, and the

upper end made to fit into the socket of a stone rest. For tinder, there
is the down from seeds of plants, impregnated with charcoal. In some
settlements, the Eskimo smoker makes fire with a couple of pieces of
iron pyrites, which they call fire stone, or flints, brought from Cape
Lisburne and the Romanzoff Mountains by traders.

The smoking process is brief and dramatic. Having set fire to a
sliver of wood, the smoker touches it to the hair and tobacco, which
go up in flames; he then pulls the whole pipeful of smoke into his
lungs with one whiff (there is no more to come) and holds it until
black spots come before his eyes, his head swims, and he is about to
suffocate. Then, he expels the air, indulges in a violent fit of cough-
ing, and refrains from trying to stand in a hurry. Some older men
(over thirty-five, that is) sometimes experience difficulty in walking
properly for several minutes after inhaling a lungful in this fashion.

So the adults, as with most adults, occasionally find the need to
escape. The children escape, too, although they do not think of it or
label it as such, in a respectable variety of games—respectable because the Eskimo is, after all, severely limited in what he can do by
lack of materials and temperature that is often thirty or more below
zero (in which, incidentally, the young play with no discomfort
whatever).

As with all children, many of their games involve imitation of
adults; the boys play at hunting and the girls at keeping house. Tag
is played as it is played anywhere. For their version of football, they
use a bag of hide stuffed with hair. They also load miniature sleds
with mouse skins and play trader; they shoot arrows, hurl spears and
small darts guided by goose feathers, and frequently hit the mark at
thirty feet. They toss each other in a walrus-hide blanket to see who
can be thrown highest into the air and still remain on his feet. Girls
have ivory dolls, often dressed in mouse or bird skins.

But most remarkable are their demonstrations of agility and dex-
terity, which, in practical fashion, provide them with amusement
when young, while teaching them the skills for survival when they
are older. They toss pebbles skillfully, juggling a half dozen in the
air at a time, with one hand; they throw stones into the air in rhythm,
catching them while crossing and criss-crossing the hands, and chant-

ing. They kick a ball of ice or snow about the size of a baseball, somewhat in soccer fashion, except that the object is to keep it in the air all the time without touching it with the hands.

In all of these things, the children reflect a characteristic good-naturedness that stems from their upbringing. Because they are doted upon, never refused anything, never whipped or reprimanded, they grow up without fear, deception, or knowing how to steal or lie. They are, in fact, highly civilized at a very early age—reasonable, extroverted and well-adjusted in their social relationships. As with their fathers and mothers, they are perpetual devotees of harmony. There is an openness, a kindness toward each other about these children, a spontaneous goodwill that approaches beauty.

There are two other matters concerning these Arctic people that are significant, and the first relates to the dog, whose value is probably best appreciated by an outsider. Lieutenant Jarvis of the *Bear* expedition wrote, "I know of no more faithful, enduring, hard-working animal than an Eskimo dog. There is no snow too deep, no ice too rough, no hill too steep for him to face and as long as there is life left in him, he will pull and struggle to drag along. Ill-fed and abused, he may seem snarling and snappish, but his faithfulness dwarfs all other considerations. For my own team, which traveled with different parts of the expedition more than twenty-five hundred miles, I have only an affectionate gratitude for the way they carried us through."

And the final matter, appropriately final, is death, which in a peripheral way is related to the dogs. These people carry their dead out on the tundra and lay them there, on top of the ground, without any ceremony except that the near relatives follow the body to its resting place. The body is usually wrapped in deerskins; if it is of a man, his sled and hunting gear are broken and laid over it; if a woman, her sewing kit and a few household goods are placed at the head, but everything so left is rendered useless.

The most interesting thing about the dead, called *nunamisimk* (which means "on the ground asleep") is that nobody talks about them afterward and to all appearances, they are soon forgotten, even by those close to them. If they were people of consequence, such as whalemen or hunters of outstanding achievement, they are not remembered so much as ordinary human creatures, but as the heroic figures of legends in which fact and fancy are blended until the tales become traditional, tend to follow patterns, and whatever the individual really was is lost sight of.

Surely they are too wise not to know what really happens to the nunamisimk left out on the hard ground—which is that the dogs, and perhaps some other animals, eat them. But they do not go there to see, and they do not talk about it, and one suspects they do not think about it, either, because there is nothing to be done about it, nor any alternative.

# 3.

# Of Arctic Life

To every thing, there is a season, and a time
to every purpose under the heaven . . .
—ECCLESIASTES 3:1

To understand what and how much an Eskimo eats is to know
how he spends the seasons and why.

An adult Eskimo eats eight pounds of meat per day and so does
each of his dogs, often eight in number. Thus, a community of
thirty people will consume more than eighteen thousand pounds of
protein in a little more than two months. Since this is roughly the
equivalent of a hundred and fifty caribou and fifteen hundred
pounds of fish, the Eskimo is necessarily in search of food most of
the time. And he labels the seasons principally in terms of what kinds
of food they make available to him, if good fortune prevails, and
the manner in which he must go after it, which hinges upon the kind
of weather. The weather also determines where he lives and in what.

In the fall, a time of shortening days and lowering temperature,
these people live in their underground winter houses, these being
holes scooped out of the ground and roofed over. Usually, they are
seven or eight feet deep and from ten to fifteen feet square, because
they are expected to accommodate two or three families. Often
about ten people live together. One family living alone is uncommon.

The framework that keeps the sides from caving in is of boards
or logs of driftwood; so is the floor. The roof is held up either by
wooden ribs or the jawbones of a bowhead whale. Over these "raf-
ters" is stretched hide, protected by a thick covering of sod. The
eaves are slightly above ground level and the roof center is a little
higher than the sides, to strengthen it against heavy snow. If ice or
snow causes a winter house to collapse, it is considered the work of

an evil spirit and no one lives in it thereafter. If the crush of ice along the shore or accumulation of snow traps and kills people inside a house, the bodies are left there, and nearby neighbors abandon their houses and move elsewhere. To some degree, this is a practical reaction, but mostly it seems to be born of fear of the hostile, unknown forces that brought about the calamity.

In the roof of the house there is a skylight, about a yard square; it is covered with strips of animal intestines sewed together and not only lets in light, but permits communication between the occupants and those outside. If an Eskimo wants to relay any news or borrow something, he hollers down the skylight, because it is a lot easier than using the chimney, which is no more than a small wooden pipe stuck up through the sod, or even the door.

The door is made unhandy on purpose, to keep out the weather. About twenty feet from the house proper, a trench is dug, the bottom of which is about three feet below the level of the hut floor. The trench is roofed over in the same fashion as the house. To enter the house, one removes a cover of wood at the entrance to the trench (taking care to replace it, once inside), walks through the trench, stoops down in front of the wall of the hut, and crawls up into the house through a three-foot hole, somewhat oval-shaped, that is cut in the floor.

Inside, on the wall opposite the entrance hole, there is a platform about three feet from the floor and it is a combination sitting room and guest room. The occupants of the hut (very often excluding the husband and wife who own it, since they may make their bed in the entranceway) sleep underneath this platform in an arrangement dictated by temperature, sex and vulnerability. On the outside lies the man; next to him, his wife, and thereafter, in a row, the children according to size, the youngest being next to the mother. They sleep under deerskin blankets, with the hair sides together. Once in bed, they remove all clothing worn during the day.

In the corner of the room, there are two or three stoves or lamps. One is stone and that is older. The others are of driftwood. But they are all in the same pattern. They are from a foot and a half to three feet long, scooped out about an inch and a half deep on top to hold the oil. Moss is placed around the edge, the roots running down into the oil, to serve as a wick. When the moss is lighted, it slowly tries the oil out of a piece of blubber that hangs a carefully figured distance above, thus replenishing the oil that is burned. The quantity of fire is controlled by the amount of moss used.

It is fall. The occupant of this house—which has been empty since last spring because the Eskimo has winter and summer dwellings—has readied it for winter, which has involved preparations both practical and ceremonial. When the place was abandoned in spring, the skylight was broken in, rubbish thrown into the house, and the entranceway blocked with snow and ice, this procedure being designed to convince the evil spirit, who might otherwise see fit to enter since there was no one to drive him away, that no one lived there anymore or intended to thereafter. By the first of October, and probably earlier, the returning occupants therefore begin their housecleaning by cutting a small chip off every piece of timber and board that can be reached easily. This is to break any spell that the evil one might have cast, even if he did think the house was forever forsaken.

Next, all inside timbers and floorboards are carefully scraped, and so are the stove-lamps.(Despite the little chimney, damper boards shifted at the entranceway, and the women's occasional waving of goose wings to shoo out smoke and odors, a lot of grease inevitably accumulates.) Ice is scooped out of the entranceway, a new skylight of walrus intestine is installed, the deerskin blankets are neatly stacked on the platform, the scaffolding outside the house (for whalebone, skins, and game) is rebuilt. The place is ready for winter living.

Fall weather is monotonously thick and stormy; it is always, at this time of year, as if this place made one last futile struggle against the tightening grip of winter—futile because the days of semi-darkness are already here and there is less and less of open ocean, until finally there is none at all.

If the Eskimo were lazy, as some have erroneously portrayed him, he would spend his winter days snug and safe in the council house, entranced by the chanting of the shaman, competing in tests of strength and agility, or telling stories about yesterday's heroes. Some of this he does, especially in the worst of the weather, but the Innuit who spends all his time socializing and thereby lets the meat supply in his ice cellar dwindle, is looked down upon. Most men go sealing in the winter.

Of all their dangerous occupations, this is the most dangerous for the Eskimos, especially since they hunt seals individually. During days that are no more than half-light, when the cold wind blows hard enough to cut the breath and the temperature ranges to thirty below, they walk miles out upon the pack—the offshore edges of which may shatter in the severe weather and go adrift with them—looking for a seal's air hole. Sometimes, a hunter is cast adrift on a floe chunk, the lead of open water between him and the shore appearing within a matter of seconds, far too quickly for him to do anything about it. In such cases, there is no warning, no surface sign of weakening ice that even the keenest eye could see—nothing but the last grim moment of split and cracking. When a hunter does not come home at the end of the day, he is never given up until the following summer, because there are men—and they are part of the legends—who have survived the winter on a drifting floe.

As important as it is to spot the mound of hoarfrost on the snow that indicates where a seal comes to breathe, the seal hunter must also keep one eye on what the pack looks like, feels like, and what it seems to be doing. For most hunters lost on the ice do not, in fact, come back.

The hunter finds a breathing hole and clears away the snow down to the hard ice. There are two ways in which he can take a seal. He can set his baleen net under the winter ice and around the air hole, scratching on the surface of the pack nearby with a seal's claws. (That will make the seal feel welcome and he will come. When he comes, he will rise to the air hole and become entangled in the net, which will drown him or break his neck when he is dragged up to the ice.) Or, the hunter can try for him with a hand spear, with a point of bone or stone. If he does this, he first drops a kind of telltale into the air hole, an instrument that works something like an ice fisherman's tip-up. It is an ivory rod about a foot long, suspended vertically in the hole by a crossbar that keeps it from dropping, and on top, there is a feather. One may sit on his three-legged stool by the air hole for a long time with nothing happening, but when the feather wiggles, the seal is coming up from below.

Since it is understood that everything the hunter kills has given its permission to be killed (or it would not have occurred at all) all hunting is closely related to specific ceremonies that must be observed, or next year, or perhaps even tomorrow, the Eskimo will not eat. So when he has killed a seal, he must get it ashore—and that may be five miles or more—in the proper manner. Sometimes it is possible to use a sled, but more often than not, he has to drag the seal to the beach.

The hand drag to be used is a two-foot-long loop of walrus hide fitted with an ivory toggle; usually, the toggle is carved to resemble two seals. The hunter may not, if he ever hopes to be successful at this task again, drag a seal with gear or line used for any other purpose, and it is, therefore, his custom to have a new drag each year to make certain that it has not been contaminated.

When the hunter comes near the shore with the seal, he removes the drag for a moment and pours a few drops of fresh water into the seal's mouth before it is taken from the ice to the land. The seal likes this, because he knows how difficult it is to get fresh water in this place in winter. And by and by, when the Eskimo has eaten the seal's flesh, hung his blubber over his lamps for light and warmth in the long night, and made summer boots and soles for his winter boots of the skin, he preserves the seal's bones and returns them to the sea, where they want to go. Other seals will know that the hunter has taken care to do this, will be grateful for it, and will allow him to kill them in due course.

In the winter, while the men are offshore for the seal, the women go after the little tom cod, which begin to run in January or slightly later. The cod school northward alongshore in tremendous numbers; the women cut holes in the ice and stand there for hours jigging the fish (that is, rhythmically jerking the line in short, rapid strokes) with unbarbed bone hooks that catch them under the chin if the fisherwoman is deft and lucky, and especially if the fish are numerous. On a good day, a fisherman may get a bushel of fish, which freeze solid as soon as they are hauled out of the water and remain that way while stored and until eaten, frozen, by the dogs usually, but by men as well if meat is scarce.

Early in the year, the people of this coast go inland for the caribou. This requires the construction of temporary shelters as they travel and of additional huts when they get into reindeer country and begin to collect the spring meat supply.

Lieutenant Ray wrote: "The snow hut of these people is very quickly and easily constructed as follows: A place where the snow is about four feet deep is selected for camp and a space five by nine feet is laid off; the upper surface is cut into blocks two feet square and eight inches thick and set on edge around the excavation, for side walls. At one end, three feet of the space is dug down to the ground or ice; in the balance, about eighteen inches of snow is left for a couch; sides and ends are built up tight and the whole is roofed with broad slabs of snow six feet thick, cut in proper dimensions to form a flat gable roof.

"Loose snow is thrown over all to chink it and, at the end, which is dug down to the ground, a hole is cut just large enough to admit a man crawling on his hands and knees; the hut is now finished; sleeping bags, provisions and lamp are passed inside, dogs are fed and turned loose, after everything they would be liable to eat or destroy is secured by caching it in the dry snow.

"After all outside work is done, everybody goes into the hut and the hole is stopped from the inside with a plug of snow that has been carefully fitted and no one is expected to go out until it is time to break camp the next morning. The combined heat from the bodies of the inmates, together with the lamp, soon raises the temperature to the freezing point and a degree of comfort is obtained that is not attainable in any other manner of camping in this region.

"The more permanent snow huts of the deer hunters, which they often occupy for a month or more, are much more elaborate. They are usually built where the snow is six or eight feet deep, so the room is high and it is approached by a covered way and an anteroom, in which the heavy outside clothing is stored. When fuel is obtainable, a fireplace is cut out of the solid walls of snow, with jambs and chimney of the same perishable material. I saw fireplaces in use that had had a fire in them for at least one hour each day for a month or more and were still intact; the parts that were exposed had softened a little under the effects of the first fire and at once hardened into ice and remained unchanged, as long as the temperature in the open air remained below zero."

When the Eskimo travels, he is very careful to respect the hostility of the atmosphere. He knows that the result of even minor carelessness can be considerable discomfort and of major carelessness, death. If he feels the slightest twinge of nose or cheek, he starts the circulation again by vigorous rubbing; otherwise, he may freeze either one and, if he does, it is painful. Usually, frostbitten skin is soon discolored, dies, and peels off, leaving the area involved sensitive for some time. If frostbite is severe, the end of the nose may drop off, or one may lose an ear (or it may hang askew, forever more).*

Now, it is March. The Innuit are ready to head back to the coast with their venison and this trek requires special preparation of the sleds.

The sleds are made from driftwood, fastened with whalebone and rawhide lashing; they are about ten feet long, two feet wide, and the runners are eight inches wide and one and one-half inches thick, straight on top and with no rail. They are shod for ordinary use with strips of bone cut from the whale's jaw and sometimes with walrus ivory, but neither of these would do for hauling a heavy load over the snow where there is no beaten trail.

So the runners are shod with ice. From the ice on a pond that is

---

* The nearer one conforms to the habits of the natives, the less likely he is to meet with disaster. In his quest for the North Pole, Peary lost his toes; his aide, Donald B. MacMillan, had frozen heels that maturated, and another member of the polar party, Captain Bob Bartlett, his cheeks black with frost, broke down on the ice pack, and called for his mother. But the Eskimo knows; especially, he knows that no part of the body requires more attention than the feet. Socks and boots must be well made and kept thoroughly dry; even the slightest perspiration will, if one stops too long, work disastrously.

free from fracture, the Eskimo cuts pieces the length of the sled runner, eight inches thick and ten inches wide. Into these blocks of ice are cut grooves deep enough to receive the sled runner up to the beam. The sled is carefully fitted into the groove and secured by pouring in water, a little at a time, and allowing it to freeze. Great care is taken in this part of the operation, for if more than a few drops of water are applied at a time, the slab of ice will be split and the work must be done all over again.

After the ice is firmly secured, the sled is turned bottom up and the ice shoe is carefully rounded with a knife and then smoothed by wetting the naked hand and passing it over the surface until the latter becomes perfectly glazed.

When the sled is ready for use, it weighs more than three hundred pounds and it is loaded with the carcasses of from seven to nine caribou, each weighing more than one hundred pounds. Men, women and children harness themselves in with the dogs to haul these loads to the coast, often a distance of more than one hundred miles. They seldom make more than eight or ten miles each day.

Once back on the coast, with daylight returned and the deer meat stored in the ice cellars, these people of the Arctic begin their seasonal preparations, both practical and ceremonial, for bowhead whaling. The council house is the center of the action each day; here, the boatheaders gather with their crews to get the gear in shape. The frame of the whale boat (*umiak*) is repaired and refastened; the old cover of sealskin is removed and replaced with a new one. Woodwork of the harpoons and lances is carefully scraped. The emphasis here is on newness; it is fundamental that the *umiak* have a new cover, that the whale cannot be approached except by those wearing new clothes that have not been previously used in hunting.

While the men are preparing their sealskin pokes (used as floats or buoys on the harpoon lines) and boat talismans, which are often eagle skins or wolf skulls, the women make the new clothing. This is in two layers, an inner suit of caribou skin, with the fat and sinew left on, for water resistance and warmth, and an outer parka of either seal or caribou skin.

Men too old for the boat crews stand daily watch on the beach, watching first for the northward-flying snowbird, whose passage signals the bowhead's coming.

The old men, the shaman among them, stand on the shore and talk loudly and often in unison, imploring the good spirit to send an east wind that will drive the ice off the shore and open a lead for the whales. The longer it takes for the east wind to come, the more obvious it is that the spirit has been offended in some manner.

Increasingly, ceremonial restrictions are placed upon the boat crews as the time for going out on the ice approaches. Their food is brought to the council house for them by the women and children; sometimes, the males sleep there, rather than going home at night, and they are not allowed to have sexual intercourse. In the last few days before the quest for whales begins, they are obliged to sit quietly together, to think only about whales and their successful capture, while the shaman chants an invitation to the northbound bowhead.

Finally, there comes the day when the whales are heard breathing in the offshore ice and finally a lead opens, often five to ten miles off the land. Because the ice in late March or early April is usually lumpy and thrown up in heaps and piles, it may be necessary for all hands to build what amounts to a road out to the lead. The *umiaks* are then loaded on sleds and hauled out to the open water by the boat crews. (No dogs are allowed on the ice in the whaling season.)

Arrived at the lead, the *umealiq* wets his harpoon in the water, sets up his talismans on the ice near the boat, and all of the crew gets into the *umiak*, which is still not launched. Then while the *umealiq* sings about pursuing and killing the bowhead, his crew simulates vigorous paddling and they go through the entire pantomime, ending when the harpooner makes one savage thrust—theoretically, the telling blow—into the empty air. Then the lines, harpoons, pokes and lances are arranged in the *umiak*; it is launched, the *umealiq* sings to invite the whale, and it is hauled out on the ice again—having been baptized for the season, as it were.

Now it is necessary to set up the camp in which the crew will live. No tents are ever set up or allowed; only the most primitive of

Hunting the caribou. *Courtesy of The Whaling Museum, New Bedford, Mass.*

Chasing the bowhead. *Courtesy of The Whaling Museum, New Bedford, Mass.*

windbreaks. No fires are permitted, and there is, therefore, no cooking done and not much eating. The boatheader, who may not allow the whale to see him eating, accordingly does not face south when he chews his meat, for that is where the bowhead is coming from. The *umealiq* covers his face with his shirt when he eats, as an extra precaution.

Then one day, there is a whale, come up for air in the lead.

One must understand the kind of courage that is required—admittedly, prompted by the desire to survive—to launch an assault against this bowhead. The land, and whatever help it might offer, is miles away; tedious and hard miles over which no great speed may be made. If the *umiak* is smashed or capsized, it is reasonable to suppose that everyone in it will die, either of injury from the whale's thrashing or of quick paralysis from the icy water. Having conceded these realities, the Eskimo paddles as hard and as deeply as he can and gets as close as he can to this black-skinned animal of thirty to sixty tons, which lies, blowing easily, and so far undisturbed.

The boats head for him in company. In the first approach, he is struck with two harpoons, each bearing a line to which are attached three inflated sealskin pokes. These floats will make it more difficult for him to sound, or dive; as he begins to bleed, their resistance in the water will tire him, and if he is killed, they will keep the carcass afloat.

The whale sounds, but neither deeply nor for long, for he is hurt; he acts tired and confused, and his efforts to rid himself of the harpoons, which have struck home hard, begin to take his strength. If unencumbered, or if the toggle-heads had been driven less deeply into him, he could have made the ice edge easily and gone beneath it and thus, beyond pursuit. But it is not to be; he needs air again, and abruptly rises to the surface in the narrow lead. Even while he blows, not easily this time, but in pain and fright, he is struck with another harpoon and his blood flows, dark upon the sea.

Now, a boat crew, sensing that the bowhead will not be able to drag down the additional floats, paddles the *umiak* in close. This is the moment of truth. If the whale is weary or hurt enough to hold still, he is likely to die. If, in his agony, he flails, it may be the men who die. Close alongside now, almost close enough to touch the shiny black skin called *muktuk*, which the Eskimo would rather eat, either raw or boiled, than anything else.

The *umealiq*, stone-headed lance drawn back, is nearly on top of the whale. There is no time left, and this thrust has to be in the heart, lungs or brain. It has to kill, and quickly. It is the heart, and in an instant, even as the *umiak* backs down quickly to get just beyond range of the flukes' last desperate smack upon the water, the vapor from the bowhead's blowholes turns red, unbelievably red against the white of the Arctic.

It is over. The whale rolls, fin out, and he is dead. In the boats, they yell in triumph, for his death is their life and every man among them remembers other springs when they took no whales at all, because a lead was late in opening, the whales had already passed, and their bellies were as empty as their ice cellars.

They drive a harpoon into his fat lip and make fast the towlines to its shaft, and all the boat crews begin the slow pull back to the ice edge. Once there, because he is not a large whale, all hands heave on the carcass until it slides out on the ice; if the whale were larger, this would have to be done piecemeal, beginning by hauling up the head, removing the whalebone, and chopping the head free of the body.

Ashore, the villagers work and watch each day during the whaling season (which may last from two weeks to two months, depending on the ice) with one eye on the offshore, for a runner will bring them the first news of a whale kill. In the meantime, all work is avoided that will make any noise, such as pounding or chopping, and no work of any kind is done in the dwelling of a whaleboat crew member. If his whaling clothes require repair during this period, the crew member obviously has to have the repairs made; but even then, his wife retires to a skin tent inland, out of sight of the ocean, to perform this task in solitude. Any contrary conduct would offend the whales and they would not come.

On this day, there comes the runner, first no more than a tiny dark accent against the sawtoothed ice horizon, yet undoubtedly every- one ashore was aware of his coming while he was still miles away, for that scrambling little man-figure was immediately translated into tons of meat, some for everybody. Yet because the whale may not be offended, the messenger's arrival, reception, and all that follows must be done according to ritual.

He bears the tip of the whale's right flipper, carries it to the wife of the *umealiq* credited with the kill, whom he informs that a bowhead has been taken, and they eat some of the meat that he has brought. It is she, dressed in new sealskin clothing prepared especially for this ceremony, who leads the villagers out over the ice road to the whale. She carries an axe or a long-handled knife and a wooden cylinder of fresh water.

Spencer relates that the traditional ceremony called for her to cut off the whale's snout, making cuts to include the snout, eyes and blowholes, at which point, the whole was stood up on the ice. From a seal-flipper pouch, the *umealiq*'s wife poured fresh water into the wooden vessel, then poured this water on snout and blowhole, saying, "It is good that you are come to us." Then the *umealiq* also poured fresh water on the snout, adding, "Here is water; you will want to drink. Next spring, come back to our boat."

Now the whole community can fall to, each member cutting off as much meat and *muktuk* as he can carry ashore, but taking care to leave the heart and flippers for the boatheader who killed the whale. As the butchering proceeds, sleds are dragged out to the site (but not pulled by dogs) and the bowhead is "taken down," everything—bone, flesh, baleen, and blubber—being usable for building, fuel, food or summer trading with the inlanders, to whom the oil was especially valuable.

Each day brings rapid change in the weather. By the latter part of April or the first part of May, snow houses are getting too soft and winter huts too damp, for comfort. It is the time of the *tupek*, the conical skin tent on a wood or whalebone frame, that is the Eskimo's "summer house." These are set up alongshore, near where the sandspits reach to the sea and the low-flying eider may be felled with a sling, near where the cold-water streams come to the sea and the whitefish may be caught in gill nets made from sinews of the reindeer.

Some few may remain in the winter villages, mostly the older people and those men who live by making things for others because they do not hunt or fish well. But generally, whole families move into the *tupeks* and the cold-weather huts lie empty. As the weather turns warmer, the whaling crews offshore are less vigilant. They may take another whale, yet the likelihood dwindles with each passing day and because they know it, they cruise less and less. Often, they haul their boats up on the ice and wait for a whale to come to them, if he is so inclined; they take turns standing watch, the off-watch sleeping or killing ducks, which flock to the open leads increasingly.

By the middle of June, all the whaleboats are brought to the land and the whaling season (since the bowhead's later migration southward is usually too late in the year and too far offshore to count on for food supply) is over for the year. The boat crews have had their feasts of boiled meat in the council house (a whale is butchered just like a cow and prime cuts come from the same relative areas). The poor have been fed. The ice cellars are stacked to the roof with chunks of the dark-red flesh. The whale hunting is finished for now.

Once the boats are ashore, Lieutenant Ray recalled, ". . . parties are made up to go to Nigalek, at the mouth of the Colville River, where the people from Nuwuk and Uglaamie (the coastal people) go to meet a band called Nu-na-ta-n-meun (inland people), where they barter oil and blubber for deer, fox and wolverine skins."

This meeting, which usually lasts for about two months, is not for the extreme poor who, having no oil for trade, scatter through the interior, carrying their kayaks on their heads to cross the numerous lakes and rivers in search of a precarious livelihood gained from catching young reindeer or fish. But for those who have something to exchange, it is a social, as well as an economic adventure, an exciting, continuing marketplace very likely unduplicated anywhere in the world.

It is customary, in this coming together, to acquire a trading partner, one with whom you deal each year. Although shrewdness is
admired, and you certainly may not become an *umealiq* without it,
implicit in the partnership is the understanding that you will make
every effort to acquire whatever it was your partner expressed a
desire for last year, that you will offer him the best you have in that
line, that you will make every effort to please him. By the same
token, he is under unspoken obligation to give you fair value for the
item, and if he believes that you have made your best effort to please,
to refrain from any show of criticism or displeasure if he is disappointed by what you have to offer.

Because all of this is so, trading partnerships usually go on for
many years; it is quite likely that they start, not so much because a
man has something you want, but because you find agreeable what
he looks like and says. They continue, often to the death of one of
the principals, because, liking each other, one's opposite number ex-
tends his best effort to find something one wants. Besting somebody
in this kind of arrangement is frowned upon. Pleasing somebody
(admittedly, while coming out of the bargain with what is wanted
and needed) is much approved and will add to one's social stature.
This is a civilized concept indeed.

These, then, are the Innuit—the people of the Arctic.

But "to everything, there is a season," and so, about the fifteenth
of August in 1871, the people of the Arctic ended their trading
and slowly began their return along the coast to their winter villages
between Point Belcher and Wainwright Inlet, hunting and fishing
along the way. They reached their winter homes about the middle
of September and bustled about to ready them for the long night.

They discovered immediately that this year was not like other
years—not like any other year.

# II

# Children of the Light

# 4.

# Old Dartmouth

We found also a great fish, called a grampus,
dead on the sands. They in the shallop found
two of them also, in the bottom of the Bay,
dead in like sort . . . They would have yielded
a great deale of oil if there had been time and
means to have taken it . . .
—Mourt's Relation

New Bedford, Massachusetts, originally called Old Dartmouth is,
of course, an entirely different matter from the Arctic. At the time
of its unspoiled beginnings and even up to the period with which
we are concerned in 1871, it was a pleasant place to look upon, and
benign in atmosphere and weather.

Eighteen years before the Pilgrims came, Bartholomew Gosnold
sailed to this site and found it "the goodliest continent" he had ever
seen, "promising more by far than we any way did expect, for it is
replenished with fair fields and with fragrant flowers, also meadows
and hedged in with stately groves, being furnished also with pleasant
brooks and beautified with two main rivers," of which the principle
one was called Cusenagg by the Indians and eventually Acushnet by
the white men.

The land east and west of the Acushnet is a matter of gentle slopes
and few hills of consequence. Its rivers run north and south; its harbors open on the south, therefore, and the rivers themselves are not
much in comparison with the Hudson or Susquehanna but are far
more modest, short-term matters, rising as brooks and streams in the
immediate countryside, deepening decently in the brief run to the
coast, and finally broadening in the estuaries so as to be navigable in
varying degrees and to provide shelter, even for deepwater craft.

At the south face of this land there is Buzzards Bay, named for the cranes or bustards that roosted on its shores for generations. Buzzards Bay is locked in on the north by the mainland, open at the east and west ends and bordered on the south by the string of Elizabeth Islands, the collection being named for an Englishwoman, but each island bearing an Indian name that inspired the old-timers to chant: "Naushon, Nonamesset, Onkatonka and Wepecket; Nashawena, Pesquinese, Cuttyhunk and Penequese," a classic example of coexistence in nomenclature.

The land is full of rocks, some strewn by the glacier's foot, long ago in the days of the hairy elephant, some parts of New England's backbone ledge that breaks the surface occasionally in monstrous slabs and even reaches out under the mud bottom of the bay. But the soil is good and black; it encourages life.

Weather here is reasonable. It can get down to zero in winter, but that is rare, and such weather never lasts; from November to March, it is as likely to rain as snow, and in summer, temperatures are sometimes in the eighties, but much more often in the seventies. There is a pleasing variety of sun, rain, wind, and fog, loosely cyclical, and seldom extreme. On most afternoons in spring, summer and fall, the southwesterly wind rises faithfully about 2 P.M. and smokes up the bay until sunset, at which time it subsides in sensible deference to the calm that day's end is expected to bring in an orderly environment.

From the northeast, especially in the fall, there come the "line storms," traditionally three days in length; they bring wind and rain, but also warm air off the ocean, so that their wet buffeting encourages the blackbird to stay in its hedge and the rabbit in its burrow, still snug and not much out of comfort.

The only cold wind is from the northwest; it comes roaring down off the Pole two or three dozen days a year and will find every blessed chink in loose shingle or puttyless pane. But it too gives up soon. And there is fog, basic to the natural idiom here; sometimes for days it lies on the horizon in a yellow bank higher by far than the Wall of China; it stays there during all the hours of sunshine, and when daylight starts to go, the whole quiet business slides inshore over the cooling earth, blotting out every landmark and winding its way inshore in white rags draped among the wet black branches of the swamps and lowlands.

Here are trees. Not such trees as stagger the senses with their size. Yet these stands are of great variety and have produced a good many ten-inch boards. In the swamps, cedar; on the uplands, the limber white birch; acres of hard, straight-grained oak; elms and maples rampant, and everywhere, the pine, which may be anything from a clear, sky-high stick inshore to a hunchbacked pile of needles that holds down a dune and refuses to let it dissolve to leeward in a gale.

Here are bearing bushes and plants. Hedging the open grasslands are head-high clumps of blueberry and huckleberry bushes; in fall, the low country blazes with carpets of cranberries and the leaves of the powdery blue beach plum; elderberry bushes droop with clusters of shiny fruit; sweet wild strawberries run the rocks, box berries brighten the forest floor.

And nuts—hickory nuts, hazelnuts; acorns and horse chestnuts. And apples, pink, yellow, flecked and red; sweet, sour, crisp and soft —and quinces, pears, cherries, and grapes, white and purple, in a half-dozen varieties. This is a land that will grow things.

On the upper slopes, there are quail, squirrels, pheasant, foxes, raccoons, skunks, deer (a few), and hare. In the marshes, seeking the wild rice, are black-necked geese, swan, and a dozen kinds of ducks, from redhead and canvasback to bluebill, mallard and teal. Offshore, the coot and eider bob in rafts, and alongshore, the saw-billed shelldrake dives in the autumn water for his dinner.

Where the cold-water streams pour down to the bay, herring, shad and smelt come in spring, fleets of flashing silver bellies, bound for the upper reaches to spawn. At the bay edge, the mud is filled with the fat, soft-shelled clam; the firm-footed quahog in his blue-rimmed shell; where the water lisps on weeded rocks, there are purple-black mussels by the colony, and crabs and lobsters scuttle across the bottom by the thousands.

From season to season, the bay is riffled with schools of mackerel and bluefish; deeper down are the rainbowed scup, the big-mouthed sea bass, and the schools of tautog.

So it is that no one and nothing goes hungry here, for everything that is natural about this place encourages and sustains life.

Yet this does not even begin to describe the natural aspects of this land which do, after all, have great bearing on the state of mind of anyone who lives here. As winter is short, spring, summer and fall are leisurely, comforting, and colorful. They produce such things to look at, to smell, to savor, as roses dripped over piles of green brambles, wild vineyards so old and thick as to form a roof of broad leaves, yellow lilies and orange poppies, blue flag in clumps, and scatterings of buttercups, violets, swamp ferns, and dandelions flung straight up the slope of a meadow to the skyline.

The shape of the land is female; it does not crag, it arcs. The shapes of the trees, because they are perpetually wind-influenced, are not Gothic, but Byzantine. Color, in all seasons, is everywhere— against the snow, the needles of pine and fir glisten dark green; in spring, the rangy forsythia pours yellow over everything; there are dozens of colors to marsh grass alone, depending on the light, hour, and season, ranging from straw to russet, and back. And always the changing sea, gray, green, or blue, depending on sky, depth, and salinity.

This, then, is a place which encourages getting up in the morning, which provides the stuff of which energy is made, as well as material to work with, whatever you may want to do. It is a land that encourages, virtually demands, growth and change once man is added to the equation.

So it was that, thirty-two years after the Plymouth landing, Captain Myles Standish, John Cook, a man named Howland and others, got a bargain when they purchased, in thirty-four shares, all of this land immediately east and west of the Acushnet River, including its rivers, creeks, meadows, necks and islands "from the sea upward to go so high that the English may not be annoyed by the hunting of the Indians, in any sort of their cattle."

They paid for this seventeenth-century Eden thirty yards of cloth, eight moose skins, twenty-two pounds in wampum, an iron kettle, and sundry other items, with which the Indian Wesamequen and his son, Wamsutta—two splendid men—apparently were satisfied.

From the beginning, the hospitable characteristics of the countryside encouraged them to undertake those things that tended to make each year better than the last. They built weirs along the shore, grist and sawmills and boatyards on the rivers. They pried rocks out of the ground and made rambling stone walls of them to contain their livestock; they plowed and then planted the fields they had cleared. They became conveniently amphibious—farmers and fishermen, carpenters and mariners—and grew to know as much about making plaster, splitting laths and shingles, as steaming and bending new frames for smack boats. Their broad-throated chimneys rose confidently against the deep forest behind. And, of course, early on, they thought about inshore whaling and dabbled in it, increasingly.

Inch by inch, these people pushed back the wilderness and established a civilization. By 1686, they had built the first town meeting house "twenty-four feet long, sixteen feet wide, nine feet stud, covered with long shingles, enclosed with planks and clapboards, with an under floor laid, benched around, with a table suitable to the length of said house, and with two light windows." Government was accompanied by law and order; in the same year, John Russell, Sr., was authorized to build a pair of stocks "forthwith," for which he was paid four and twenty shillings, and by 1709, Henry Howland built another pair and a whipping post as well.

A workhouse was established in 1742 for "the setting to work of all idle persons," and a bounty of twenty shillings was paid for every wolf killed within the township and three shillings for every grown wildcat killed in town from "the last day of September to the first day of March and six shillings for each wildcat killed the other part of the year." Daniel Shepherd was chosen schoolmaster and given "eighteen pounds and his debt" for a year's service.

It was still a hamlet, awaiting its future. The architect of that destiny—a man with energy, ideas, and money—proved to be Joseph Rotch of Nantucket. He came to New Bedford to set himself up in the whaling business, which was at that time a matter of six-week to two-month coastal voyages in relatively small vessels of from forty to sixty tons.

Business attracts business and thus were established in this place Benjamin Taber, block-maker and whaleboat builder; Joseph Russell and John Loudon, shipbuilders; Gideon Mosher, mechanic, and Elnathan Sampson, blacksmith, and a handful of others, whose names and skills remained trademarks here for generations and whose accumulated wealth has impact in this community even today.

It was they who prosecuted shipbuilding with vigor, built wharves, opened roads, encouraged the multiplication of shops and houses, and the flourishing of a coastal fishery and merchant service. New Bedford vessels began to reach out, their owners sensing the opportunity that a wider world of effort offered. Under a grove of buttonwood trees by the bank of the Acushnet, the keel of the village's first big ship—the *Dartmouth*—was laid; she was owned and built by Joseph Rotch. On the December night in 1773 when the Sons of Liberty threw the tea overboard in Boston Harbor, Rotch's ship *Dartmouth* was one of the tea fleet.

By late 1792, Dartmouth (now called New Bedford and the name Dartmouth thereafter applied only to the adjacent town to the west) had its first newspaper, the *Marine Journal*, published by John Spooner, near Rotch's Wharf. Mr. Spooner sold books, including Bibles, testaments, and hymnbooks, "any of which will be exchanged for clean cotton, linen rags, old sailcloth, or junk." His advertisers included Joseph Clement, compass maker; William Rotch, Jr., who dealt in sailcloth, and cordage, and Joseph Damon, who, on November 27 of that year, offered for sale "a large, well-built vessel, just launched."

Six years later, the contents of a second newspaper, the *Columbian Courier*, revealed that New Bedford was not only bustling but possessed a sophisticated, cosmopolitan population. Caleb Greene and Sons were dispensing drugs and medicines; Peleg Howland, European and West India goods; Jeremiah Mayhew, dry goods, carpets, china and crockery ware—the dry goods were European imports.

True, wars intervened occasionally and slowed the community's growth. During the American Revolution, Britain's Sir Henry Clinton "directed Major-General Gray to proceed to [New] Bedford, a noted rendezvous for privateers, etc. and in which there were a number of captured ships at the time." General Gray's force, aboard thirty-six transports, escorted by two forty-gun frigates and an eighteen-gun brig, numbered four thousand light infantry, grenadiers, and regiments of foot. "Bedford Village," as it was then called, had one detachment of artillery with a single gun and "fewer than fifteen able-bodied men in the town."

The British marched down King Street to the waterfront and burned a total of seventy vessels in the river, including eight prizes and six privateers. They also destroyed by fire twenty-six storehouses containing rum, sugar, molasses, coffee, tobacco, cotton, tea, gunpowder, sailcloth and cordage, and further burned a distillery and two rope walks, the former providing several minor explosions, and the whole business making a large light in the night for some hours. Wrecks of the burned vessels remained in the river mud for years and were a navigational bother.

And in the War of 1812, which New Bedford opposed bitterly, knowing its far-flung vessels would be fair game for the enemy (in the presidential election of 1812, its citizens cast 399 votes for Clinton of the Federal or peace party, and 13 for Madison, reelected by the Republican or war party), the port was closed against all traffic —an economic disaster—and a British brig patrolled just outside to make sure no vessel went in or out. In the first three months following the declaration of war, New Bedford had eight vessels captured by the enemy. With cargo, they were worth $218,000.

Yet as serious as these setbacks were, they did not halt New Bedford's advance more than temporarily.

Matters went swimmingly in most of the first half of the nineteenth century. And then came the Civil War, which was uniquely regarded by an important segment of the New Bedford population. As anticipated, it provoked economic disaster. Yet the war's underlying cause, the federal effort to abolish slavery, had strong support in the community. Not among the workingmen—it is a matter of record that shipyard hands refused to work alongside a black, even though it was the mayor's wish—but among its wealthy leaders, many of whom were Quakers.

The tradition of abolitionism here was deeply rooted. As far back as 1770, Elnathan Sampson had purchased at public auction in Dartmouth a black male named Venter, age forty-six, for twenty-one pounds, six shillings and five pence and thereupon "acquitted and renounced all Right, Title or Interest whatever in and to said Negro" and "set him at full Liberty to act his own will from the day of the Date hereof forever."

The cost of the Civil War to New Bedford was high. The whaling fleet was swept from the seas; of the forty-six vessels destroyed by the rebel cruiser *Alabama*, twenty-five were from this port. The ships lost were valued at $1,150,000 and the oil aboard at $500,000.

In 1860, New Bedford's valuation was $24,196,138; in 1865, it was $20,525,790; the decrease was due principally to the effect of the war on the whaling industry compounded by the emergence of fuel substitutes for oil.

Local historian Leonard Bolles Ellis supplied a realistic economic assessment of New Bedford's situation at the war's end:

". . . Our idle wharves were fringed with dismantled ships. Cargoes of oil covered with seaweed were stowed in the sheds and along the riverfront, waiting for a satisfactory market that never came. Every returning whaler increased the depression. Voyages that in former times would have netted handsome returns to owners and crews resulted only in loss to the one and meager returns to the hardy mariners. Such was the condition of affairs when peace came in 1865.

". . . it was clear that something must be done to save the city from a permanent decline. The natural advantages of climate and situation for the development of cotton manufacture and of kindred operations were seen and capital, of which the city had an abundance, was soon finding rapid and profitable investment in home industries."

All of this notwithstanding, some did stay with whaling—including George, Jr., and Matthew Howland. The building of the bark *Concordia* was proof enough of that.

# 5.

# Two Quaker Brothers

[George] Fox loved to dwell on the light of
Christ. "Believe in the Light, that ye may become children of the Light." was his message
again and again. So much did he and his followers dwell on this, that though at first they
called themselves "Children of Truth," they
were soon termed "Children of Light," a name
which they adopted and used for some time.
—CLAPP AND THOMAS,
*A History of the Friends in America*

George Fox was the founder of the Society of Friends, called by
some, Quakers. The latter term is misleading; it was first used by
Justice Bennet of Derby, England, simply because Mr. Fox had bade
his followers tremble at the word of the Lord.

It is more useful to remember that Mr. Fox said, "When the Lord
sent me forth into the world, He forbade me to put off my hat to
any, high or low." And Mr. Fox also said, "I used in my dealings
the word 'verily,' and it was a common saying among people that
knew me, if George says 'verily,' there is no altering him."

Dartmouth, later New Bedford, arose because there was no altering the followers of George Fox in the New World.

A considerable number of references relating to the difficulties of
the seventeenth-century Quakers with their Puritan peers may be
found; in these items, the depth and persistence of the conflict becomes immediately apparent and because of it, Quakers left the
Plymouth Colony and were principal among those who founded
Dartmouth, which became New Bedford.

The roots of their differences were many and complex, but the
formal structure of the Society of Friends of that time demonstrates
both its debt and opposition to orthodox Puritanism. Undeniably,

the Society was a closed one, a self-righteous, self-supporting con-
venticle that recognized no higher temporal or spiritual authority
than the immanence of God within each of its members. It imposed
rigid standards of speech, dress, and demeanor, and allowed no
deviations whatever. It demanded absolute loyalty, total militancy.
But unlike the Puritans, who believed their exclusions were a right
and a privilege, the Society of Friends saw exclusions as an obligation
and an opportunity.

The Puritans sat in judgment of the world. The Quakers judged
themselves and hoped the world would follow suit. Thus, in a very
substantial way, the Society of Friends was the first active proponent
of peaceful coexistence. It served itself to better serve mankind.

On the matter of materialism, too, the Quakers broke with the
Puritans in a significant way. Like the Puritans, they undoubtedly
believed that worldly success was an implicit sign of Election, probably the most reliable, maybe the only outward sign. It was no accident that the principal oligarchs of the Plymouth Colony were its
principal landowners. Or vice versa. Or that religious orthodoxy
tended to maintain itself on property lines, increasing as it approached the economic heart of the colony, decreasing with the
distance from it.

Under the heading of "Trade," a nineteenth-century publication
on discipline and advices of the Society of Friends reminds: "Being
earnestly concerned that the service of our religious Society may
not be obstructed, or its reputation dishonored by any imprudence
of its members in their worldly engagements, we recommend to all
that they be careful not to venture upon such business as they do
not well understand, nor launch out in trade beyond their abilities,
and at the risk of others, especially on the credit which may be
derived from a profession of the Truth; but that they bound their
engagements by their means . . ."

The Society membership was urged to "be careful how they enter
into joint obligations with others under the plea of rendering acts of
kindness; many by so doing have been suddenly ruined and their
families reduced to deplorable circumstances."

However, unlike the Puritans, the Quakers were conscience-bound to justify this material proof of God's bounty in terms of the common good. Perhaps the Quaker definition of the "common good" was a narrow one, yet it went beyond anything that the Puritans recognized. Small wonder that the Puritans were anxious to chase, or at least squeeze, the Quakers from the fold.

But what gave the Quakers their ultimate singularity (as well as their nickname) was a kind of simplistic emotional commitment to God that made their lives on earth a permanent love affair. Love was George Fox's answer to all human problems. "Suffer any wrong that can be done you without retaliation; never requite evil with evil." "Swear not at all." "In stillness, there is fullness. In fullness, there is nothingness. In nothingness, all things." As love presumes equality, equality presumes love. As love is strength, strength is love. Love——God's inner Light. God's loving children. Children of the Light.

Quakers have been defined by some as Calvinist humanists. Calvinists, necessarily, because they derived so much of their outward character from Calvinism and built from the same limited theological material. Humanists surely, because they affirmed the beauty of life and the right of all mankind to progress and benefit by God's example. They believed that love, not terror, might redeem the world. If, as Oliver Wendell Holmes said of Jonathan Edwards, the Puritans "drove a nail through the human heart," the Quakers wanted to remove the nail and dress the wound. By their Light, the Quakers would be both charitable and compassionate. The Puritans, by conviction, would be neither.

So the Quakers left the Plymouth Colony without hesitation or regret to pursue the inner Light. Thus it happened that the township of Dartmouth received many of these people and, in due course, they became a strong element in the community, outweighing in influence and outvoting that portion of the inhabitants who were in sympathy with the government at Plymouth.

As a result, Dartmouth was constantly in trouble with the government because of refusal to support the religious service insisted upon by the Court. Both Quakers and Baptists resisted taxes urged by the Court for the building of meetinghouses and for the maintanence of a ministry whose creed did not appeal to their sense of truth.

When, in 1675, King Philip, king of the Pokanoket Indians, led more than a thousand warriors in a raid on Dartmouth, murdered several men and women and burned about thirty houses there—an attack stemming principally from the white man's inhumanity and impudence—Plymouth, which was responsible for deterioration of relationships with the Indian, had its own theological idea as to why the Dartmouth tragedy occurred.

In an order passed October 14, 1675, the Plymouth Court noted: "This Court, taking into their serious consideration the tremendous dispensations of God toward the people of Dartmouth in suffering the barbarous heathen to spoil and destroy most of their habitations, the enemy being greatly advantaged thereunto by the unsettled way of living, do, therefore, order that in the rebuilding and resettling thereof that they so order it as to live compact together, at least in each village, as they may be in a capacity to defend themselves from the assault of an enemy and the better to attend the public worship of God and the ministry of the word of God, whose carelessness to obtain and to attend unto we fear may have been a provocation of God thus to chastise their contempt of his gospel, which we earnestly desire the people of that place may seriously consider of, lay to heart, and be humbled for, with a solicitous endeavor after a reformation thereof, by a vigorous putting forth to obtain an able, faithful dispenser of the word of God among them and to encourage them therein, the neglect whereof this Court, as they may and must, God willing, they will not permit for the future."

The essence of this statement seems to be that if the people of Dartmouth had had a minister acceptable to Plymouth (and built their houses closer together), Philip would have left them alone. Philip's reaction to this, regrettably unobtained and unobtainable, would have been interesting.

Dartmouth's reaction is more easily come by. In the records of the Society of Friends, it is related that "At a mans [sic] meeting in the Town of Dartmouth the: 6 Day of the 11 month 1698/9 at the house of John Lapham, wee underwritten Peleg Slocum, Jacob Mott, Abraham Tucker, and John Tucker, the day and year above written, undertakes to build a meeting house [at Apponegansett in Dartmouth] for the people of God in Scorn Called Quakers 35 foot long, 30 foot wide and 14 foot studds to worship and serve the true and living God in according as they are persuaded in Contience they Ought to Do and for no other use, Interest or Purpose but as aforesd . . ."

Ten years later, Nathaniel Howland and four other members of the Dartmouth Monthly Meeting were impressed for military service in Canada. They refused to serve and were taken before Governor Dudley at Roxbury and given a hearing and discharge. Two years later, Nicholas Lapham and John Tucker, Jr., members of the same society, refused to enter military service and were imprisoned four weeks and two days in the Castle at Boston.

As early as 1772, the Dartmouth Monthly Meeting deplored the practice of holding their fellowmen in bondage and appointed a committee to visit the few members of the Society who were still slaveholders and "labor with them in such manner as the case demanded." Three members held six Negroes "in bondage," declined to free them, and were accordingly labeled offenders in the matter by the committee. The Society majority eventually prevailed, however, and approximately ten years later, none of its members any longer held slaves.

The American Revolution and the years leading up to it marked a painful time for the Dartmouth Quaker; if religious differences strained the ties with Plymouth, military differences did the same with his neighbors to the east in Fairhaven.

A classic example occurred only twenty-four days after the battle of Lexington. A couple of dozen enthusiasts from Fairhaven set sail in an old sloop named *Success* (armed with one unworkable carriage gun that went overboard the first time it was fired) and recaptured two Yankee vessels that had been seized by British naval forces in the area. Several British prisoners were taken.

The New Bedford Quakers were disturbed by this act, not only because they were opposed to war in principle, and because a state of war hardly existed at that time, but because they were afraid of a British counterattack by the British sloop of war *Falcon*, Captain Linzee, which was only a few miles away in Tarpaulin Cove on Naushon Island.

Accordingly, a Quaker delegation from New Bedford went to Fairhaven resolved to return the prisoners and captured property to Linzee with an apology. The intention was frustrated not only because the men of Fairhaven had already divided the spoils and marched the prisoners off to a jail more than twenty miles away, but further, the Quaker effort was not looked upon by its critics as having been born exclusively of lofty motive.

As one of those who participated in the *Success* venture recalled, "At New Bedford, a large majority of its influential citizens were of the Society of Friends, by principle and profession noncombatant and as they had large commercial interests afloat and exposed, it was quite natural that this outcropping of belligerent patriotism by their neighbors across the harbor should excite in them, as it did, an earnest feeling of repression."

The Quaker, then, was, in many instances, persecuted, maltreated, and discriminated against. In turn, his public demeanor was intensely provocative; it was often characterized by ridicule of religious observances, obstruction of the enforcement of ordinances, disturbing of public meetings and rejection of broadly accepted customs, the whole being—in the eyes of his opponents—virtual evidence of religious and social revolution.

Yet above all, the Quakers were the mainspring of the community. Life with them was occasionally abrasive to others, yet, peace and war notwithstanding, matters forged ahead—exciting, irrepressible, determined, and principally civilized and economically triumphant, led—or rather, "pushed," to use Edmund Burke's term—by the energetic, opinionated, self-contained and tireless Friends.

A principal "pusher" was George Howland, Quaker. Since his heirs, George, Jr., and Matthew, who built *Concordia*, had much to do with the whaling industry for more than half a century, to know something about the first George Howland is to understand what the industry was and especially the manner of its demise in his sons' time.

Old George, the son of a Fairhaven farmer, died in 1852, aged

George Howland, Jr

Matthew Howland

seventy-one years, leaving a net estate of $615,000, a fleet of nine whaling vessels, a countinghouse, wharf and candle factory in New Bedford; acreage in Maine, western New York State, Michigan and Illinois; a wholly nominal title to Howland Island in the mid-Pacific, and charitable bequests in the amount of $70,000, of which $50,000 went toward the foundation of Howland College in Union Springs, New York; $15,000 to help found Haverford College in Pennsylvania, and $5,000 for various other Quaker causes.

The important thing to remember is that George Howland's financial success was duplicated forty or fifty times over in New Bedford during the course of his lifetime. In short, New Bedford, in the mid-nineteenth century, was perhaps the richest city per capita in the world, and its rise to riches was breathtaking. In 1800, the total population was 4,361 and the community's income from the whale fishery was probably less than $300,000. By 1830, its population was up to 7,592, but its annual whaling income was already close to $3,500,000. Ten years later, with a population of 12,087, its whaling fleet grossed $7,230,000. In 1854, with a population of barely 20,000, the city imported, principally with its own fleet, $10,802,594 worth of whale and sperm oil and whalebone.

In the same year, the city's first textile mill, the Wamsutta, increased its capital of $600,000 and had thirty thousand spindles and six hundred looms in operation. Add the gross business of the community's candle manufacturers and allied industry, and the gross product approximated $20 million annually.

This prosperity came about, in part, because from 1825 to 1875, American whalers had the world's whaling grounds virtually to themselves. Of an estimated nine hundred whaleships of all nations engaged in whaling in the late 1840s, more than seven hundred were American; as late as the 1850s, from four hundred to five hundred of these were from New England.

The year that George Howland died, his fleet comprised nine vessels, the *George, George Howland, St. George, George and Susan* (obviously, the name George held his attention somewhat), *Onward, Java, Golconda, Corinthian,* and *Rousseau.* Their book value was roughly $300,000, their replacement value perhaps $800,000 to

$1,000,000, and it is unlikely they could be duplicated today for less than $7 million—$10 million, if you made a first-class job of it.

This fleet was, therefore, a considerable legacy for George Howland, Jr., and his half-brother, Matthew.

George Howland, Jr., was the only surviving child of his father's first marriage. He was born in 1806, educated at Friends Academy in New Bedford and in Germantown, and by French tutors in New York; he entered the whaling business with his father in 1820, aged fourteen. In addition to being handsome, poised and ascetic, young George was a good man in the best sense of that word; he was good where his father had been worthy, good in the spirit of his convictions where his father had been good to the letter of them.

This seems fair to say, for surely it was no easier to combine the demands of business with those of active public service in the latter half of the nineteenth century than it is in the latter half of the twentieth, yet George, Jr., did an outstanding job of it.

He was, among other things, a member of the corporation of the New Bedford Institution for Savings, president of the New Bedford Five Cents Savings Bank, director of the New Bedford and Taunton Railroad, treasurer of the New England Yearly Meeting of Friends, Whig representative to the General Court of Massachusetts, selectman of the town of New Bedford, five-term mayor of New Bedford, from 1855 to 1865, after it became a city; state senator, member of the Governor's Council of Massachusetts, founder and first benefactor of the New Bedford Free Public Library (to which he gave his first two years' salary as mayor, or $1,600); trustee of the Taunton Lunatic Hospital, fellow of the American Society of Engineers, member of the Executive Committee on Indian Affairs for President Grant, a manager of Haverford College, and a trustee of Howland College and Brown University.

George led, Matthew followed. One wonders whether this was, in part, because George set out to be his father as he thought his father should have been—a conscience-driven, second generation acceptance of noblesse oblige. On the other hand, the fact that Matthew followed was, to a degree, relative and inevitable. Someone had to tend the counting room while George pursued the public interest. Someone had to post the cashbooks, oversee the candle works, cover drafts, hire, fire, and ride herd on the whaling fleet. And George, with his eight years' seniority had, in effect, put in his time in the whaling business already; he was clearly now aimed for other, higher things.

George was dedicated, likable, intelligent, and sensitive. During a journey in Italy, he went to Civita Vecchia, arriving there on February fifth, the last day of the traditional carnival. He wrote, "We were exceedingly amused in witnessing the sports and in which we joined to some little extent . . . At about half past five, they commenced the final portion of the amusements of the day by having in their hands or on sticks small wax candles, of which there must have been thousands.

"The sport consisted of blowing out those within reach and when they had succeeded in extinguishing all within a carriage or window, they would set up a tremendous shout of '*smoccola, smoccola.*'

"I could not help entering into this for half an hour or so with as much zeal as any of them. In fact, I exchanged bunches of flowers and put out lights with people whom, of course, I never saw before and probably shall never see again."

Everyone respected George. The New Bedford *Republican Standard*, when he was first sent to the General Court, observed that he had been voted for by "the colored abolitionists." But the Civil War posed a particular problem for him, since he was the city's mayor at the time. The Society of Friends, of which he was an active, lifelong member, could not, consistently with its doctrines, give approval to the war. But he felt that since the Union could not endure without a bitter struggle, a vigorous prosecution of the war was plainly a necessity. His conception of the duty of a patriotic citizen once formed, he threw all of his energy and influence as the community's chief executive into the encouragement of recruiting and attention to the welfare of the departing troops.

Draft riots occurred in other Northern cities. It was feared they might occur in New Bedford and this was a difficult prospect for a Friend, who could be called upon at any moment to give orders that would result in bloodshed. But George's concept of duty was clear; he garrisoned the City Hall and established mounted patrols to guard the roads leading into New Bedford, with the aim of intercepting those who might come seeking to instigate riots.

His sister-in-law, Rachel, Matthew's wife, was a foremost Quaker minister. She asked George what he intended to do in case the feared insurrection should occur in New Bedford. He replied, "There will be no blank cartridges fired."

As representative of the President of the United States, George spent several weeks with the Osage Indians in Kansas, compiling what was described as "a fund of information that proved highly serviceable in determining what was needed to bring about a desired improvement in their mode of living." Unfortunately, his personal correspondence on this subject has disappeared; one wonders what he had to say about the plight of the Osage (assuredly, he was, as always, both empathetic and practical) and what, if anything, ever happened to his recommendations.

George's belief in New Bedford and its future was limitless and unshakable. In his address on September 14, 1864, on the occasion of the two hundredth anniversary of the incorporation of the Town of Dartmouth (of which New Bedford had initially been a part), he declared, "When I look over our city, and see the improvements which have taken place within my time, and over the territory represented by you, my fellow citizens and neighbors, and then go further, and embrace the whole country, I sometimes ask myself the question, 'Can these improvements continue? And will science and art make the same rapid strides for the next fifty or one hundred years . . . ?'

"The only answer I can make is the real Yankee one; why not?"

George's personal and family experience with whaling—wars, shipwrecks, and market vacillations notwithstanding—undoubtedly buttressed his faith in what was to come. In seven voyages from 1829 to 1851, ranging from twenty-four to forty-four months duration, the Howland whaleship *Golconda* produced a gross return of $334,345. In eight voyages of from thirty-seven to fifty-three months, from 1838 to 1866, the comparable figure for the *George Howland* was $583,020. Of this $917,365 total, the crews were paid

25 percent or about $228,665. Of the remainder, $455,000 was paid by the owners for outfitting expenses. Estimated owners' earnings left amounted to $233,700.

Annual earnings of just these two ships ranged from $3,000 to $5,000; figuring original investments of about $25,000 per vessel, their profits earned yielded an investor return of from 13 to 21 percent annually over thirty years.

One of the most successful vessels of the Howland fleet (and for the longest period) was the *Rousseau*, built in 1800 and bought by the family at a Philadelphia auction—George, Jr., went to bid on her—about thirty years later. When the ship arrived in New Bedford, the first thing that old George did was to have the figurehead of the "French infidel" for whom she was named chopped off and thrown into the mud of the dock. He was much irritated when informed that the vessel's name could not be changed (somebody misinformed him, for ship's names were, in fact, changed), purposely mispronounced it "Russ-o" all of his life and consistently poohpoohed her excellent catches.

However, he did not return her to her former owner in Philadelphia, although he once threatened to do so. Presumably this was because, under Howland ownership for the forty-four-year period beginning in 1834, *Rousseau* produced more than eighteen thousand barrels of oil and thirty-nine thousand pounds of whalebone; on seven out of twelve trips, she took more than one thousand barrels of sperm oil and on one trip, more than two thousand. These figures stand as exceptional to this day.

Both George, Jr., and Matthew undoubtedly sensed that the New Bedford whaling fleet was overextended, possibly as early as 1856. But, among other factors encouraging to them, they had an advantage over latecomers to the whale fishery. Most of their vessels had paid for themselves five and six times over and were in the hands of unusually competent, purse-shrewd masters. This meant that any serious break in market prices or demand would affect the Howland fleet last, while less solvent firms were sure to go under.

Surely the brothers did not welcome the softening market, yet if it pared the number of their competitors, it had a positive use. Further, although in 1858, forty-four out of sixty-eight returning whalers had losing voyages, this did not mean that all of the statistics were bad for everybody.

Three Howland ships came home in 1866. The *Corinthian* and *George Howland* docked in April with a total of 930 barrels of sperm oil and 8,100 barrels of whale oil. A few months later, the *Onward* discharged sixty-two thousand pounds of whalebone on the Howland wharf. The gross value of these three voyages was $620,000, on a capital investment of less than $90,000. It is reasonable to assume that George and Matthew received of this total at least $150,000 apiece. And no income tax.

True, some kind of petroleum product was a long-term threat to the whaling industry, but in the meantime, something financially titillating had occurred. This was the emergence of a new kind of whaling, only eighteen years old—pursuit of the bowhead in the Arctic seas, the bowhead being a so-called whalebone whale.

Even while oil prices fell, because of heavy stocks on hand, whalebone was quoted in 1855 at forty-five and a quarter cents a pound and by 1858, at ninety-six and three-quarter cents, a rise of more than 100 percent. The bone, strong and flexible, capable of being shaped, and relatively light in weight, was much used for such items as corset stays and buggy whips. The bowhead has from 514 to 630 slabs of whalebone, depending on which authority does the counting. One bowhead is recorded as yielding bone weighing three thousand pounds; another made 375 barrels of oil. The bowhead is a heavy yielder and the rule of thumb is that for every barrel of oil taken from him, there will be seven to seventeen pounds of bone.

These figures were something to think about. George and Matthew thought about them. Any reasonable man, and they were certainly reasonable, knew that Arctic whaling gave a shipowner no more than a gambler's chance. A ship could easily get crushed in the ice or, if it were a "closed" season because of consistently low temperatures, solid ice pack would prevent the ship from getting to the whaling grounds. Still, there was always the chance that a vessel might make a catch worth as much as a hundred thousand dollars in a summer's work. And that also was something to think about.

The Howlands considered a large fleet the cheapest form of insurance. It had worked well for the first George and now, with premiums running upward of 10 percent of insured valuations, they felt they really had no choice. In the best of all worlds, of course, they would not have sent their ships to risk the perils of the Arctic whaling grounds. But since it was increasingly the Arctic or nothing, they decided to send their ships north. If they lost a vessel in the process, so be it. It was worth the risk and cheaper than insuring.

From the cost standpoint, it was also helpful that no member of a whaleship's crew, from the captain down, received fixed wages. Each man on board was paid a certain share, or lay, of the net result. Commonly, a captain received a lay ranging from a tenth to a fifteenth; what he got depended on how successful he had been in previous voyages and, therefore, how much in demand he was. The three mates, first, second, and third, received from an eighteenth to a forty-fifth. Foremast hands got anywhere from a one hundred and fiftieth to a two hundredth. But from the owner's point of view, the important thing was that if the ship took no oil or bone, or if she sank, burned, or was wrecked, nobody owed officers or crew anything. Everybody signed on for a percentage of the catch. No catch, no percentage.

And finally, bowheading was more like shooting fish in a barrel than any other hunting the whaleman did. There was no cruising the world's wide seas for years on end, as in sperm whaling; no fighting a vicious whale, with resultant casualties, smashed equipment, and lost time, for the bowhead killed easily, and no wasting effort on a poor whale, for all the bowheads were fat.

In terms of mileage, man hours and outfitting costs, bowheading represented a minimum effort for a possible maximum return. The season was short, roughly from sometime in July to the middle of September; the area of hunting was small, involving a triangle in the Chukchi Sea north of Bering Strait and to North Cape and Blossom Shoal on the west; east to Point Barrow or, at most, one hundred miles east of that, and south to East Cape. The current in this sea, flowing northward from the strait, splits to the east and west and a perpetual tongue of ice—pointed south—is thus formed, marking the northern limit of navigable water. The whalers called the tongue "Post Office Point." It wasn't always in precisely the same place, but it was always there as a handy reference mark.

The Arctic whaling fleet sailed perpetually between the land and the offshore ice, in some seasons working up the Siberian side, in others, cruising to the eastward of Point Barrow if ice permitted; the distances they sailed were relatively short. In general, they figured that a bowhead was worth $10,000; four big ones comprised a good trip. In stubborn seasons, when the ice hung on, the open-water areas in which they could operate were even smaller than usual. They sailed much together, worked the same areas at the same time, and except for foggy periods, usually were in sight of each other, which they preferred because the possibility of trouble with the ice was always present.

Bearing all the factors of Arctic whaling in mind, George and Matthew decided to add a tenth vessel to their fleet. This vessel was *Concordia*. And if, in retrospect, this grand expenditure appears to have defied all the economic storm signals, it may help to consider it in the total context of the Quaker's self-defined role—as epitomized by the Howland brothers—in the community.

These Quakers assumed a responsibility for economic, political, and moral leadership, part of this involving an unflagging faith in the future; they were aggressive shapers and molders, and were busy at this task, in major and minor ways in a half-dozen avenues simultaneously, all of the time.

When a library was donated to New Bedford by the Friends Society in 1813, a committee of members went over the list of books and discarded many, such as several foremost English poets and Shakespeare's works, as unfit for young people to read. This opposition to certain aspects of culture and the arts, implemented by the influential Quaker leadership, was dominant in the New Bedford area for many years. On February 2, 1857, George Vandenhoff, temporarily retired from the stage, came to New Bedford to present an evening's entertainment before the Lyceum. George Howland, Jr., sat on the platform as president of the Lyceum society.

Mr. Vandenhoff presented a series of readings and in one of the selections became somewhat dramatic, whereupon George Howland interrupted him, declaring, "This is not a theater!" Mr. Vandenhoff responded by reciting the reply of Jacques to the Duke in "As You Like It," beginning, "All the world's a stage . . ." The audience applauded but it is recorded that "the Quaker element present saw no sense in the laughter."

Because the rise and strength of the Society of Friends had evolved simultaneously in New Bedford and Nantucket, born of parallel factors and comparable people, an incident occurred in 1867 —the year that the bark *Concordia* was launched—which more objective observers might have found ominous.

As clerk of the New Bedford Monthly Meeting, Matthew Howland entered in the records an excerpt from the Nantucket Monthly Meeting of Men and Women Friends, which read: "Our clerk is directed to deliver our records and papers to the clerk of New Bedford Monthly Meeting and furnish that meeting with a list of members as far as can be ascertained.

"With this minute is closed the existence of Nantucket Monthly Meeting, which commenced under the ministry of Thomas Story about the year 1704. Through the influence of Divine Grace abundantly vouchsafed to the inhabitants of this island, it increased until its membership included a large portion of our citizens, but by the operation of adverse circumstances and a dividing Spirit, it is reduced to a mere handful."

A blow surely, a grievous occurrence. Yet if the New Bedford Quakers privately thought it a portent of anything, neither their records nor reactions betrayed this. On the contrary, they responded in characteristic fashion, proposing to encourage and assist the holding of some meetings on Nantucket, with the aim of restimulating interest, and, in the short run, trying to keep such island membership as remained within the fold. George was named one of a committee of four to "report their judgment on a list of [Nantucket] persons who it appears are entitled to the right of membership."

Efforts of the committee were successful in part. But where they were not, there were clear signs of new-mindedness that some might have found disturbing. Three weeks after *Concordia* was launched and while she was still being outfitted at the Howland wharf in New Bedford, George Howland's committee delivered one of its several reports to the Monthly Meeting on attempts to round up the wayward. "We have again had an interview with Edward Mitchell. He says he sees no cause to change his former decision and he will, therefore, remain with Fair Street friends [this was written with a lower-case "f"], the 'seperate [sic] body.' We believe that further care on our part will not be useful."

And, invited to become a member of the New Bedford Monthly Meeting, Nantucketer Mary Allen replied, "I am a member of the Nantucket Monthly Meeting of Friends and always have been, and no other." At least that kind of response gave some hope that a Monthly Meeting on the island might be reestablished, given encouragement and support.

There still were evidences of vigor in the New Bedford society. At this time, it was operating several "first-day" schools in the area—the one in New Bedford had 249 members, of whom forty-four were Friends. Moreover, there may have been some conscious effort among the membership in the direction of slightly more liberalism aimed at reducing organizational fractures for lesser reasons.

But what the Quakers were slow to realize, being too close to the matter, was the degree to which their theological destinies were linked to the whaling industry. Even conceding the degree to which their ranks were diminished by doctrinal differences, it remained for Quaker historians of a later generation to observe that "the failure of the whale fisheries of Nantucket and New Bedford led to a very general exodus" of Friends.

By 1850, there were fewer than five hundred orthodox Quakers in the New Bedford society; by 1887, their minutes noted, "the whole number of members now on our records is 346."

One trouble was that, whatever whaling did by way of financial ups and downs, it posed a problem for them. If it was outstandingly successful, as it had been for some, they accumulated a lot of money; since it is a truth of some kind that economic prosperity is, over the long run, inimical to a plain style of living and since they could hardly give away *all* their assets for good works, they had to decide between finding theologically acceptable ways of spending it or getting out of the Society.

Some tried to spend their extra money on more or bigger and better houses; in some cases, the very kind of houses to which their fathers would have objected on grounds of ostentation. There was some excuse for a bigger house if you had a lot of children but even that did not explain to the disapproving orthodoxy why there was an increasing Quaker movement west from the modest houses along the Acushnet River up to the more pretentious neighborhood of "The Hill," which was looked upon as a Unitarian stronghold.

Matthew Howland had a summer place at Clark's Point in the south end of the community (which he called "The Cottage") and justified it as a practical matter by planting part of the land and raising produce for the family's use.

Still, there were a lot of people with a lot of money and some of them, with an eye on the whaling slump, scattered westward. The country was opening up at a furious rate; capital was much in demand and the rates of return were fantastic. Here, surely, was a chief cause of the "very general exodus" of Friends. The very prudence and perception that had caused them to sense the opportunity in Old Dartmouth and to exploit it successfully drove them to migrate to areas of new opportunity.

And, on the other hand, if whaling was not successful, and it had not been for some, the resultant financial adversities diminished their power base and influence. For although numbers were certainly important to the Society of Friends, their impact upon New Bedford—the achievement of a tightly disciplined theocratic society—arose because of who they were, rather than how many. Their successes in government, their management and direction of community affairs for generations were linked inextricably to their successes in investing, banking, trade, local economic development, and to the extensive philanthropies they were able to afford.

Finally, some did not migrate to the West, did not falter in their dedication to the Monthly Meeting, and did not forsake whaling, either. George and Matthew Howland were among those who did not. For them, the whale fisheries had not yet failed; more importantly, their natures were not such as to command a break with orthodoxy or removal from a community with whose destiny they had been so closely married. George had been far more active in civic matters than had Matthew. Yet the Howland brothers were among the first thought of when something had to be done—a meeting chaired, an occasion observed, a political figure endorsed, funds raised for good works—and surely both sensed to what degree New Bedford was what they had decided it ought to be. Sentiment apart, if a relationship has gone on long enough, there comes a time when it is easier to find a bearable compromise than to end it, even if it is not what it ought to be or used to be. George and Matthew were men to whom the possibility of bearable compromise existed and to whom change was unpalatable; they simply kept on doing what they were doing.

George certainly, and Matthew very likely, even though he was slightly younger, must also have been aware that the average age of the Monthly Meeting member was rising; they were losing by death about nine or ten members each year and the youth of the community had noticeably less patience with the society's disciplines, and eyes and thoughts in other theological directions.

Still, George and Matthew kept on doing what they were doing. George led and Matthew followed.

# 6.

# "A Ramble in New Bedford"

> The town itself is perhaps the dearest place
> to live in, in all New England. It is a land of
> oil, true enough; but not like Canaan, a land also
> of corn and wine. The streets do not run with
> milk, nor in the springtime do they pave them
> with fresh eggs; yet in spite of this, nowhere in
> all America will you find more patrician-like
> houses, parks and gardens more opulent than
> in New Bedford. Whence came they? . . . all
> these brave houses and flowery gardens came
> from the Atlantic, Pacific, and Indian Oceans.
> One and all, they were harpooned and dragged
> up hither from the bottom of the sea.
> —MELVILLE, *Moby Dick*

In 1871, Edward King, correspondent of the Boston *Journal*, visited the bank of the Acushnet River to write a couple of pieces entitled "A Ramble in New Bedford." What he discovered and concluded follows here. The parenthetical interjections are from the New Bedford *Republican Standard* of the same period since, although Mr. King did a creditable job, he could not, of course, learn of, or include, everything.

I first interviewed Captain Edmund Gardner, a massively formed, venerable gentleman with no sign of infirmity and, with the subdued polished manners of half a century ago, he rose to meet me. "What does thee want?" was the natural question after introduction and the good captain had sunk back into his armchair in the window recess.

(*Potomska Mills—At a meeting of the stockholders of the mills, it was voted to further increase the capital stock by the issuing of*

*2,000 shares at $100 each, making the whole stock $500,000. The stockholders are to have the exclusive right to take stock for thirty days to two-thirds the amount they already hold. We are glad to see this evidence of a growing interest in the manufacturing business.*)

Captain Gardner bears some marks of his adventures, a whale having playfully taken him in her mouth when he, as master, was urging on his crew to trap the monster fish. One hand was crushed and the whale's toothprint may be seen in it, while a deep skull wound attests to the power of the monarch of the seas. "I turned over in her mouth," said he, "and saved myself that way."

(*Watchman Wesley Furlong arrested two men who were fighting near the corner of Middle and Purchase Streets. One of them knocked him down and kicked him, and they both escaped. Quite a number of fellows watched the proceedings, but though called upon for aid, would not assist the officer.*)

Captain Gardner is eighty-seven years old and he has been retired from the sea nearly a half century. If he is the product of the Friends, they have good reason to be proud of him. He is an owner in the ship *Roman* and is one of the many venerable captains in New Bedford who have seen the sun set in every region of the globe and braved every conceivable danger at sea.

(*What shall girls who work for a living do? In this city, we are told there are a lot of shopgirls who are working hard for barely enough to pay their board. This is what ought not to be. Girls ought not, as they do, to feel themselves above entering upon domestic employment. We can remember when the wives of some of our most opulent merchants were so engaged and we will wager that neither they nor their husbands regret the fact to this day. Good domestic help is always in demand; especially is there a demand for American help. The worst result of the system that has prevailed in regard to female employment has been to fill the state with marriageable females who know no more about household duties than an Esquimaux and who are, therefore, as unfit for wives of industrious merchants as could possibly be imagined. God designed women should be helpmates for men and they should exercise a stronger influence over their minds than has been the fashion of late years. They should cease to look upon marriage only as the means of greater extravagance and a wide scope of pleasures and upon a husband only as a golden goose who pays the bills.*—Letter to the Editor, signed Ike Marble)

It is very difficult, however, to make any of these captains admit there is comparatively any more danger at sea than on land. Most of them, indeed, it would be difficult to name the exceptions, are pious, God-fearing men and as strong physically as morally.

(*The Memorial Day oration was delivered by the Reverend O. A. Roberts in clear and lively tones. Referring to our dead, he said, "Here are some of the noble fifteen hundred who sailed from this port, embarked in the floating castles of our Navy, to exchange the harpoon for the sword, who sailed with Farragut, all-conquering on the Father of Waters; with Porter, leveling the fortifications on the coast, or with Winslow, consigning the* Alabama *to her long-deserved fate—some of you know of Monocacy, with its fearful odds, and of the waves of defeat and victory that swept the Shenandoah Valley, culminating in Early's annihilation at Fishers' Hill. You remember the stonewall of Antietam, where Massachusetts and New York blood together flowed, but not in vain; the circumvention of Port Hudson, its burning sands, oft-repeated charges; Gettysburg, with its all-decisive vantage ground; and Rewara, where our Massachusetts comrades fell dead into the arms of the foe.*")

Vessels used to leave the mouth of the Acushnet with their crews singing that good whaling song, the oldest extant, one verse of which ran, "When in our stations we are placed, whales around us play; We launch our boats into the main, and swiftly chase our prey."

(*J. L. Sisson, at his fish market, corner of Purchase and Hillman Streets, received this morning twenty striped bass weighing over five hundred pounds, caught by his brother, Otis A. Sisson, at Nomansland with rod and reel on Monday afternoon and Tuesday morning. We don't hear of any of Sir Isaak's disciples doing any better. They are a very handsome lot of fish.*)

The only variety of costume one sees nowadays in New Bedford is exhibited by the Quaker, many of whom habitually wear the wide, broad-brimmed hat, gray clothes and irreproachable necktie

Union and Fourth (now Purchase) Street, New Bedford, c. 1870.
*Courtesy of The Whaling Museum, New Bedford, Mass.*

of that order. There is hardly an hour of the day when that costume may not be seen here and it is inseparably associated with wealth and honorable old age.

(*The colored people of this city are planning a celebration of the great events of their history, to take place on the eighteenth of July, in which they can make use of such funds as the community may see fit to grant them for the purpose.*)

New Bedford has a cosmopolitan breeze always blowing over its social strata; although its waterside streets are not as picturesque as those of Bordeaux, one hears the same strange mélange of languages there as in the French city. There, on that vessel fitting out for its long and adventurous cruise, you may hear all the modern languages spoken.

(*It is gratifying to notice the many improvements that are being made by our people in their dwellings in several localities. Never before perhaps has the work of "brushing up" been carried on to such an extent. One cannot ride through the city without being pleased at the evidence of thrift and comfort which prevails. Particularly is this the case in the northwest portion of the thickly settled part of the city. Strangers should not judge our people by the appearance of the old buildings on the lower streets of the city, but rather by the residences, imposing or otherwise, with few exceptions, which will be found on a line west of Purchase Street in the north, and west of Second Street in the south. Our city has the credit of being the best governed in the state, if not in the nation, and we challenge comparison also in respect to beauty and cleanliness. We know of no other city that offers so many inducements for a place of residence.*)

There is every type of feature in a stroll down Water Street; your eye has to be well put there and alert to distinguish the rising generation, for they are gradually becoming Americanized, but if you look sharply, you will distinguish Portuguese and French descent in the smart, black-haired girl with the olive complexion, tripping schoolward with her handful of books.

(*Where are the patriotism and public spirit of New Bedford? Where are the promises made by capitalists at "war meetings" to*

*induce young men to enter the national service? Where are the sym-
pathies of the community for the needs of those made needy by the
exigencies of war? Where are the citizens who have thronged the
entertainments of Post 1, G.A.R., in previous years? We are ashamed
to say that the drama of "The War for the Union," for the benefit
of the Grand Army Charity Fund, was given again on Saturday
evening to a small audience. When people can be pleasantly entertained by a large dramatic company, comprising several fine artists,
and a host of amateur talent, with music by a full band and orchestra,
the hall should be crowded nightly.)*

New Bedford has a quarter known as Fayal, so many Portuguese
inhabit it. Dutchmen, Frenchmen, Spaniards, Norwegians, and
Scotch have also pitched their tents here, and make periodic visits
to whale regions. They are a prosperous and very generally a peace-
able lot; the Portuguese, all along the coast, have an especially good
reputation, not only as brave and hardy seamen, but as orderly
citizens.

*(The temperance meeting was well-attended. The Rev. A. Parke
Burgess of Boston spoke and he called this the best temperance city
in the world. The Rev. L. B. Bates of Taunton said New Bedford
was the only city of twenty thousand inhabitants in the world where
a prohibitory law is in force. He said in twenty years of his ministry,
he had attended funerals of fifty men who died of intemperance,
not one of whom had passed his twenty-fifth year. He said that
twenty-three hundred a year died from that cause in Massachusetts
and predicts that if the sale of ale becomes general, the number will
be increased another thousand. He said he believed in praying and
voting as the most efficient means of carrying on the temperance.)*

"You may marvel at our general air of quiet," said a prominent
citizen to me. "I tell you we have always been a lazy community,
sending our ships away, waiting for them to come in, and living
quietly; the town was slow to wake from the kind of lethargy into
which it had fallen, but business is reviving now, the character of
it is somewhat changing, manufacturing is coming in, and the Irish
with it, and we shall have lost many of our hitherto distinctive
characteristics."

The intersection of Elm and North Water streets, looking north.
The countinghouse of G. and M. Howland was on the corner of North
Water and North streets. *Courtesy of The Whaling Museum, New
Bedford, Mass.*

(*The meeting of the trustees of The Five Cents Savings Bank on Saturday last was an anniversary, it being just one year since they occupied their new banking room. The report of the treasurer of the bank showed the bank to be in a very flourishing state. They congratulate themselves on the change of location and confess that in that, as in all business, advertising pays. The deposits are $2,277,-723.49, an increase of deposits for the past six months of $294,148. Accounts opened the past six months, 1,197. Accounts closed the past six months, 200. Accounts now open, 9,708. They have opened an average of 200 a month for this year. The amount of deposits for the year has been $758,009.79, an increase of $284,696.87 over last year's work. Opening the bank for business, the treasurer thinks, on Saturday evening, has induced many a young man to lay away something against a time of need from his week's earning. On last Saturday evening, there were 100 entries. The management of the bank is in excellent hands, aiming to invest the funds in their care with regard to safety, instead of large dividends.*)

A tendency to conservatism has heretofore marked the dealings of New Bedford with the outside world, but it is gradually giving way before the influx of manufacturing. Cotton mills now monopolize the north and south ends of the town; iron and copper works employ large numbers of men; steam manufactures are attracting flocks of young folks, whose heads are filled with anything but Puritan notions. So New Bedford is in transition and this condition of affairs is puzzling to the venerable captains and retired merchants, whenever they venture outside their cozy mansions into the "city."

(*There is certainly no objection to the dealing out of medicine on the Sabbath, but it does not follow because this is allowed that the Sabbath should be made a holiday in every apothecary shop and that young men should congregate there to drink and smoke through the day and evening. Our apothecaries ought to be men to whom a reasonable appeal could be made successfully. They should not wait for the strong arm of the law to compel them to respect the Sabbath or the citizens who desire to enjoy it as a day of rest and religious improvement. We hope that they may look at this matter and see what is its bearing on the morals and good order of society.*—Letter to the Editor, signed E.)

There are elms in all the streets and, in midsummer, when clad in garments of living green certainly must produce a lovely effect. But it is cold here now and I climbed a roof just at sunset and looked down upon the town. Under the glare of the dying sun, it looked quite weird as it scrambled up the hillside, as if anxious to escape from the river and the little forest of masts.

(*The mob who seized upon Paris and drove out the Thiers government, establishing a Communist one in its stead, have met with no sympathy from the friends of true liberty anywhere. Their composition appears to have been very much that of the band who went out to join David in the wilderness, ". . . everyone that was in distress, and everyone that was in debt, and everyone that was discontented, gathered themselves together, to the number of four hundred and followed him." The leaders of the Paris insurgents are obscure men; their acts are violent and cruel; they abolished the claims due for rents; they set up the Revolutionary calendar; they cruelly murdered their fellow citizens; they threatened to set up the guillotine; they talked of disposing of all the national properties, etc., all in the name of "liberty, equality and fraternity," under which the French have perpetrated so many cruelties and barbarities. Such is not the way to set up a truly free government, nor to establish liberty on a firm foundation. The civil war is a horrible thing. Everyone will rejoice that they have been defeated and hope that the people of France will, in general, sustain wiser men and better counsels.*)

The Acushnet River does good service; the river wharf frontage is very extensive and solidly built; vessels are constantly coming and going, with freight for and from the iron, copper, glass and oil refining works. In 1871, an average year, there were 62 arrivals of vessels from foreign ports, having a tonnage of 17,906 tons, and employing nearly 1,600 men. The estimated value of the product of the whale fishery brought into port in 1871 was $2,503,962. The ships brought 1,083,105 gallons of crude sperm oil, valued at $1,375,-

402; 1,614,832 gallons of crude whale oil, valued at $954,210, as well as 368,433 pounds of whalebone worth $264,164.

(*In the Congress, they are still debating on the Ku Klux Klan. Gen. Butler gave what was described as a forcible speech in asking for the remedy for the trouble of the South. He said that if the Democratic Party of the North and South would proclaim that it is necessary for the success of the democracy that these murders, outrages and wrongs should stop, and life, property and all the rights of citizens must be respected or they could not choose a Democratic president in 1872—from that hour, in his belief, profound peace and quiet would reign in every county in the Southern states, insomuch that a hated and despised Yankee schoolmarm might teach, undisturbed, Negro children from the Potomac to the Rio Grande. Yea, even how to read the Scriptures. But he said it was evident they could not do this and there was nothing left for us but to pass strong, vigorous laws to be promptly executed by a firm hand, armed, when need be, with military power.*)

The ironworks have vessels constantly sailing from London to New Bedford, with cargoes of old scrap iron, and the copper and other manufactories keep a large number of vessels in operation. One thing that one learns by a visit to New Bedford is that the whaling trade is still of vast importance to the town and is of vital interest to the country. New Bedford is practically the whale-oil center of the United States. Of course, petroleum has made savage inroads on the sale of oils taken from the "monsters of the deep," but there is still a business in very large amount done here.

(*Marine news: This forenoon, the scene below Palmer's Island was an animated one. Two ships and one barque were coming up and one barque going to sea, with over twenty sloops and schooners coming in and going out.*

(*At Central and Taber's Wharves, five ships and barques are being fitted for sea.*

(*The Brig* Hallie, *Captain Jackson, which arrived at San Francisco from the Arctic Ocean, reports sighting Captain Frederick A. Barker, the first, second and fourth mates, cooper, five boat steerers,*

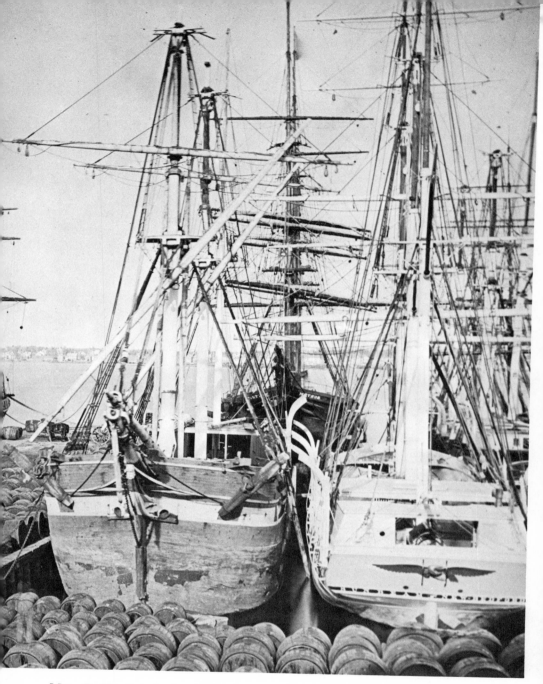

New Bedford waterfront in the 1870s. At right is the celebrated bark *Sunbeam*. The dark-hulled bark at left of wharf is a cargo vessel, not a whaler, and is, perhaps, unloading oil from Honolulu. The old New Bedford-Fairhaven bridge is just visible in the background. *Courtesy of The Whaling Museum, New Bedford, Mass.*

*and fifteen sailors of the British whaling bark* Japan, *which was wrecked on East Cape in October last. The third mate and eight men were lost. Captain Barker and the crew wintered at Plover Bay. This will be joyful news to his widowed mother and sisters, who reside in this city at 105 Maxfield Street.*)

The Friends demanded the same sternness of morals at sea as on land from all who went out in their ships. When the *Rebecca*, the ship on which the famous Captain Cornelius Grinnell sailed, was launched in 1785, the handsome female figurehead which the builders had given her was sternly ordered by the owners to be cast off as a vanity. Some of the young men held a funeral over the figurehead on the shore, as the vessel departed.

(*Young Rogues Arrested: Captain Perry yesterday arrested James Johnson, thirteen years old, and Roland F. Jenney, fifteen, who were charged with breaking into Charles H. Beetle's store on Thursday night last, and they have confessed to that and other similar operations. Both have been at the Farm School; Johnson, for assault with a knife and Jenney, for till-tapping. Jenney was sentenced by Judge Bennett to the School Ship. The case of his companion, Johnson, was continued. Last Friday night, in his cell, Johnson was on his knees for two hours, blubbering and whining, groaning and agonizing, slobbering and snorting, and promising never to steal again, as long as he lived, provided the Lord would get him out of this scrape.*)

Among the most considerable firms in New Bedford engaged in the business of manufacturing oil are Hastings and Company, whose works cover an area of three acres and whose wharves are four hundred feet long, with a frontage of one hundred feet. Two hundred barrels of oil are daily carried through the ordinary processes, yet only twenty men are employed. During each year, Hastings and Company purchase and refine about forty thousand barrels of whale oil, which they send all over the world. It is used mainly to mix with other oils and the demand for it is quite as great as at any time before petroleum was put into active use. This firm gives but little attention to the manufacture of candles, finding a readier market for spermacetti in cruder forms. Sales average nearly one million dollars yearly.

(*The fact that Jeff Davis is allowed to go about making speeches* of this kind is a sufficient answer to his talk about oppression. The government has by no means done its worst nor all it should have done to him personally; though there are no objections to an amnesty for the deluded people on whom Davis brought the suffering of which he complains.)

Delano and Company and W. A. Robinson and Company each have establishments covering about two acres and manufacture from 25,000 to 30,000 barrels of whale oils annually. George S. Homer and Company has a very extensive establishment devoted mainly to the refining of oil for other parties in New York and Boston and perfects many thousands of barrels yearly. Samuel Leonard and Son, established in 1837, deal extensively in sperm and whale oil and make immense numbers of candles yearly. Their business amounts to about $150,000 annually and they refine perhaps twenty-five hundred barrels of sperm and one thousand of whale oil yearly.

(Some gentlemen who were in Fall River yesterday noticed a great difference in the appearance of the streets there and here. Several of the mills had just dispersed a month's pay and during the afternoon and evening, as many as twenty men were seen drunk in the streets, some of them utterly helpless. Fighting was a frequent occurrence and the police were not around or, if visible, did not appear to notice the transactions unless they were of unusual violence. The fact is, there is considerable difference between prohibition rule and the state of things where this question does not enter into the election.)

The manufacture of the patent candle, hot-pressed and containing a large proportion of wax, is gradually becoming an important industry in New Bedford, and oil soap, a product obtained after various pressings, much demanded in woolen manufactures, is produced in large quantities. This business of oil refining has shown a great decrease in the number of manufactures in late years, but about the same amount of capital remains invested. The majority of the crude oil which the New Bedford whalers bring home is refined here. Wales and Company, manufacturers of paraffin and wax candles, will work up two tons of fine stock weekly; Thayer, Judd and Company, who, with fifteen candle machines, work up about fifteen hundred pounds of stock daily into paraffin and wax candles; the New Bedford Paraffin and Lubricating Oil Company, and the Petroleum and Coal Oil Company, established in 1860, which distills and refines three hundred barrels of petroleum per week, are in flourishing condition. All the products of petroleum, benzine, naphtha, and superior burning oil are perfected in the latter establishment.

*(Some time ago, Governor Claflin, with a view of testing the question, submitted to the Executive Council of the state the appointments of Mrs. Julia Ward Howe and Miss Stevens of Cambridge, as justices of the peace. The opinion of the Supreme Judicial Court was asked in the matter, and the unanimous opinion of the court is that, by the constitution of the Commonwealth, the office of justice of the peace is a judicial office and must be exercised by the officer in person and a woman, whether married or unmarried, cannot be appointed to such an office. The law of Massachusetts, at the time of the adoption of the constitution, the whole frame and purport of the instrument itself, and the universal understanding and unbroken practical construction of the greater part of a century afterward, all support this conclusion and are inconsistent with any other. It follows that if a woman should be formally appointed and commissioned as a justice of the peace, she would have no constitutional or legal authority to exercise any of the functions appertaining to that office.)*

The metal-working interests of the city depend to some extent upon the whaling industry for their success. The copper works make sheathing for the ships and a large manufactory is devoted to making iron ship-castings and large numbers of workmen are constantly supported by fabricating bolts, nails and machinery for whalers. New Bedford Copper Company, located in the northern section of the town, which wears more of the modern aspect than does any other quarter, was incorporated in 1860 and has a capital of $500,000; ninety workmen are constantly employed and three million pounds of copper and metal are annually used. The copper, which all comes from the Lake regions in ingots, is sent from New York to the company's own wharves at New Bedford by water.

New Bedford waterfront c. 1870. Vessel drying sails is the *Massachu-setts*, which was subsequently lost in the Arctic in 1871. *Courtesy of The Whaling Museum, New Bedford, Mass.*

The works are very extensive, comprising a rolling mill, a mixing house, and a huge refining house. The manufacture of the bronze sheathing for ships, which is so-called because of its color, requires two million pounds of metal yearly and is the source of a very large revenue for the company. The company does an immense business in the manufacture of printing rollers of copper used by manufacturers of print cloths for engraving patterns upon, and the makers are accustomed to saying with pride that they have to manufacture all the English rollers sent to this country.

(*At a meeting of riggers, held at Music Hall on Tuesday afternoon, the following statement was agreed to and it has been signed by twenty-one journeymen, all but two of the whole number in the city: We the undersigned riggers of New Bedford and vicinity, knowing our trade to be an arduous and dangerous one, and that there are only a few months in the year in which we can work at all and believing, in consideration of these facts and the high prices of living, that our wages are not unreasonable and if a reduction should be made therefrom, it would make no difference in regard to the number of ships to be refitted, agree not to work for any less than what we now receive, namely, forty cents per hour.*)

The Gosnold Iron Works, which take their name from the bold namer of the islands in the bay, pays one hundred stalwart workers $6,000 monthly and consumes six thousand tons of scrap iron and five thousand tons of coal annually. The value of its yearly products, which find their chief markets in New York and Boston, is about $300,000 yearly. Most of the workmen are foreigners and the manufacture is mainly hoop iron, bands and rods, of which fifteen tons daily are made. The rolling mill and adjuncts cover five acres and have seven steam engines and ten boilers; the scrap iron comes mainly from Belgium and Holland through London, from which port, it is brought directly to New Bedford by water.

(*New Bedford retail prices listed: Chicken, 30 cents per pound; potatoes, $1.20 to $1.50 per bushel; onions, per peck, 60 cents; apples, per peck, 70 cents; fresh cod, 10 cents a pound; coffee, 24 cents to 50 cents a pound; Franklin coal, delivered, $10 per ton; milk, 7 cents* *a quart; fresh beef, 20 to 25 cents a pound, includes sirloin, rib roast, and round; corned beef, 10 to 19 cents per pound.)*

Flour, cordage, carriages, shoes, glass, fine houses and pretty women are among the other productions of New Bedford. The manufacture of carriages has grown to be an extensive interest. William G. White and Company manufactures $10,000 worth yearly; H. H. Forbes, established in 1862, do a business of $30,000; Brownell, Ashley and Company employes twenty-three persons and has a $40,000 yearly business. Gifford and Allen, and Andrew Cragie have extensive establishments for making brass castings for ship work.

*(Whenever a remarkable murder or other crime takes place in New York, the press make a great ado about it, preaching homilies about the sad state of things in the metropolis. What they are saying is true enough. New York is mainly governed and controlled by scoundrels and ruffians. Some of them live in Fifth Avenue palaces and revel in all the luxuries that wealth can obtain. Some of them are mere rough bullies, pickpockets, thieves and would-be assassins, but they are all of one stamp and while the people waste a good many words on the state of things, they do nothing to remedy it. They are too timid or they are too busy with their wares and their merchandise to attend to their political duties. They are not willing to serve on juries or, by their united movement, to place the municipal control of the city in honest hands. They are so divided by political factions that they cannot combine to give themselves a decent government. The fault is their own and the remedy is in their own hands. If they do not do this, things will go from bad to worse until the only cure will be a movement like that in San Francisco a few years ago which, by summarily hanging a few desperadoes, cleared the city of them, and has kept it a decent city ever since.)*

Then there are the flour companies, the principal one of which, christened after the city, New Bedford Flour Company, produces 150 barrels of flour daily and disperses it all over Massachusetts and Rhode Island, selling directly to country dealers and averaging nearly $500,000 in yearly sales. A cordage company, whose extensive rope walk flourished when the British soldiers came to burn New Bedford, now does a handsome business, and one or two boiler and plane companies are in active operation.

*(Sperm oil is indispensable for the running of any kind of machinery and all who have anything to do with fine machinery eventually come back to this first principle. Many manufacturers of some machines have heretofore used some of the many substitutes, but nearly all have at last come to the inevitable conclusion that there is nothing like pure sperm oil. The putting out of sperm oil in small bottles for use on sewing machines has come to be a large business, as will readily be seen when the fact is known that it is estimated that three thousand barrels will be used in this country the present year, in the manufacture and running of sewing machines.)*

The cotton mills, north and south, old and new, will, in a few years from now, do much to transform New Bedford into an active and bustling city with perhaps all the vanities and vexations of large communities. The Wamsutta Mills now cover a large extent of territory at the north and the new Potomska mills at the south. The four mills of the Wamsutta Corporation are the largest buildings in the city, that commonly called No. 4 being five hundred feet long by seventy-six feet wide, with three stories, basement and attic. Sixteen hundred hands are at present employed; annual consumption of cotton is 10,000 bales; the capital stock has been increased to $2 million; it is nearly all owned by present or former residents of New Bedford.

The corporation proposes to build another building; ninety thousand spindles and two thousand looms are kept running, and the sales annually amount to something over $2 million. The corporation was organized in 1847, with a capital of $160,000 and put its first mill into operation in 1849. By 1865, it had three in operation.

Quite a little village has gathered around the Wamsutta Mills and it is expected that the Potomska Mills will make the southern section lively likewise.

*(Consistency: To spend $1,500 in taking care of our Common, a barren hold, and to spend $200 on all of our public cemeteries, where*

*lies the precious dust of those who have created much of the wealth of our city. To give a man $600 to look after the Common occasionally, and to pay a man $400 to fill the delicate and responsible office to deal with truant children, many of them orphans, and many others with parents that cannot control them, an office that requires as much or more wisdom and discretion than any other in the city.—* Letter to the Editor, signed Shoddy.)

Melville was, of course, correct in concluding that, in all of America, New Bedford's opulence was outstanding. The community was rich, on a per capita basis, but relatively few people, more often than not related, held most of the money. It was shrewdly governed (by the same people) with an eye toward keeping down the taxes, but most of the population did not pay property taxes anyway, because they couldn't afford to own property. It was a bustling place economically, but there were always enough people out of work so that one didn't have to pay very much to hire them. It was as enlightened as any other New England community bursting at the seams with investment capital and poised on the threshold of the manufacturing era, but most entrepreneurs saw little use in sending workingmen's children to school when they could get jobs and supplement the family income. And although there was extensive private philanthropy to supplement the state's limited effort, the almshouse was an accepted social instrument.

One political party (Republican) dominated not only government at all levels but was so exclusively the party that one *ought* to belong to, in terms of economics, social intercourse and everything else, that it was subconsciously associated with breeding, brains, affluence, and acceptable genealogy. Labor unions, Democrats, the Irish, people with unfamiliar accents or ideas and, generally, any man who was poor and thought he had a right to get rich if he could (even though this was precisely what New Bedford's founding fathers had done) were privately or even publicly looked upon as ridiculous or dangerous.

On the one hand, the leadership of the community made extensive contributions; on the other, it at least condoned an atmosphere of imposed single-mindedness that ranged from provincialism to bigotry.

These two items appeared in the principal newspaper only a few days apart, in 1871:

—The Free Public Library has just received about $5,000 worth of books purchased from the income of the Sylvia Ann Howland fund, which will add much to the collection as soon as they can be arranged for public use. The recent purchase includes forty volumes of miscellaneous pamphlets, collated, annotated and indexed by Lord Ashburton, which could not have been procured otherwise for three times their cost.

—Agents are sought for a book entitled *Romanism As It Is*, containing 750 pages and 105 first-class engravings. It fully uncovers the Roman system from its origin to the present time, exposes its baseless pretenses, its frauds, its persecutions, its gross immoralities, its opposition to our public schools, and civil and religious liberty. It shows its insidious workings which strongly tend to bring this country under full Romish control.

Along the New Bedford waterfront, 1871 was the usual busy year, its day-to-day bustle punctuated by arrivals and news dispatches from the far places with which its people were concerned:

—Captain James Hyland, master of the bark *Rousseau*, George and Matthew Howland, owner-agents, which sailed from this port last October, came home sick.

—Arrived, bark *Ohio*, Loum Snow and Son, from the Atlantic grounds, after three years, with 350 barrels sperm, 285 whale, 1,477 bone; sent home 1,100 sperm, 1,273 whale, and 30,581 bone.

—*Desdemona*, Captain Samuel F. Davis, G. and M. Howland, Atlantic grounds, sent home 50 whale, 800 pounds bone.

—William Michael, fourth mate, *Lydia*, Captain Lysander W. Gifford, Edmund Maxfield, owner-agent, Pacific Ocean, died at sea.

—Captain John P. Praro, bark *Crowninshield*, Terry and Chase of Fairhaven, owner-agent, Pacific Ocean, received Order of the Rose from the Emperor of Brazil, for saving crew of the Brazilian brig *Damao*.

—Bark *William and Henry*, Captain Daniel B. Green, Atlantic grounds, sent home 414 sperm, and condemned at Fayal.

—Whaling schooner *Montezuma*, Captain Leach, Atlantic grounds, towed into Vineyard Haven, dismantled in a gale.

—Arrived New Bedford, ship *Europa*, Captain John G. Nye, from Indian Ocean; sailed October 23, 1867, arrived July 31, 1871; F. Armstrong, 3d mate, died September, 1868; fourth mate drowned, 1869.

—J. F. Mandousa, third mate, bark *Cicero*, Captain Henry Clay, Atlantic grounds, dropped dead in his boat while fast to a whale.

Considering the nature of the whaling business, it is unlikely that any of these dispatches struck the whaling merchants of New Bedford as being exceptional or untoward. In hindsight, however, one news item in particular stands out with dreadful force. On September 26, 1871, the *Republican Standard* noted that, "the news of the loss of the bark *Oriole*, received by mail last evening from San Francisco, is not generally credited. The owners have had no news in relation to her and it probably grew out of her having been stove by ice earlier in the season, the injury not having been very serious."

This report may have been discredited at the time it first appeared. Soon enough its accuracy would be demonstrated beyond the shadow of a doubt.

# 7.

# The Countinghouse

We affectionately desire that Friends may
wait for Divine counsel in all their engagements
[of trade] and not suffer their minds to be
hurried away by an inordinate desire for
worldly riches; remembering the observation
of the Apostle in his day, so often sorrowfully
verified in ours that, "They that desire to be
rich fall into a temptation and a snare and many
foolish and hurtful lusts."
—Rules of Discipline and Advices,
Society of Friends

In the New Bedford city directory of 1871, an entry states,
"Howland, Matthew—merchant, North, near North Second; house,
37 Hawthorn Street," which is to say that Matthew Howland lived
at the top of the hill among those who had money and worked at
the bottom, where it was earned, adjacent to the waterfront where
the wharf of "G. and M. Howland" was located.

The principal preoccupation for Matthew was the brothers' business, yet in noting this, one must avoid the simplistic error of assuming, as some do, that because a man spends his days with columns of
figures, is shrewd to the penny and does things precisely the same
each time—since that is his nature—nothing of other consequence,
neither morning sky nor bird song, nor the perennial plight of man,
invades his awareness.

This is popular caricature, complete with the inevitable distor-
tions; Melville and others may have concluded that the master of the
countinghouse had ice water in his veins—and some did—but just
because there is no poetry in double-entry bookkeeping, it does not

follow that there are no bookkeepers who read poetry. Besides, Matthew was much more than a custodian of ledgers and an administrator of properties, and there were deep forces far removed from money matters constantly at work within him.

First, it could be argued reasonably that his daily tasks, while not obviously concerned with God, were never wholly divorceable from Him, in the mind of this man who grew up in a deeply religious home and who was the son of a woman who was a much esteemed minister in the Society of Friends and the husband of another. Scholarly Daniel Ricketson, whom Thoreau had visited and corresponded with over many years, had been close to Matthew since school days. Of Matthew, Ricketson said, "He accepted literally and without qualification the letter of the inspired page—to him it was the voice of God and to walk therein was his highest aim and duty," and added that he was "representative of a class of our older Quaker merchants now so nearly extinct."

Surely to Matthew, religion was a daily matter. He watched the dollars carefully. Yet it is significant that, whereas some owner-agents encouraged their masters to scrimp on food aboard ship and made excessive profits on sale of the clothes and other items in the vessels' slop chests, there is evidence to the contrary in the case of the Howland fleet in Matthew's time.

Moreover, he was sufficiently familiar with the "Advices" to remember their warning: "Even when riches to any extraordinary degree have been amassed by the successful industry of parents, how often have they proved like wings to their children, carrying them beyond the limitations of Truth into liberties repugnant to our religious testimonies . . ."

Matthew's father had "amassed riches to an extraordinary degree." Matthew was aware of these spiritual dangers, and the ever-present prick of conscience must have served as a steady governor of his life machinery. In much the same vein as that of the "Advices," he once wrote to one of his children, "Thee must remember . . . not to allow anything to creep in . . . and destroy that love and interest [in family] which is worth more than money."

Another factor was physiological. For many years, Matthew had

been a semi-invalid—Ricketson referred to the "hard and unrelent-
ing affliction, against which he had to contend through life—" and
this appears to have had two principal effects. In some of Matthew's
letters, there is the suggestion of persistent depression, which was
one of them. Yet constantly overriding this was the strongest kind
of will; he refused to let his ailment, whatever it was, inhibit in major
degree his steady round of activities embracing church, state, business and society. As a matter of fact, even with regard to his depres-
sions, one cannot know to what degree they arose from infirmity
or how much of them was simply the response of a sensitive man to
those matters in life—especially involving harsh and rapid change—
that were not agreeable and yet not easily avoidable.

In any case, there is evidence that, although Matthew avoided
flamboyance of expression and flights of fancy by nature, he was
not, in fact, prosaic. This is the man who wrote to one of his children, "I wish thee could have been with [us] last eve. After tea, we
went down to the wharf and took Will's boat and went out on the
river, sailing around the yachts, who began to send up fireworks
about 8:30 P.M. till ten. There were about forty of them, and the
river was all ablaze and looked as though it were on fire. It seemed
like faerie land. The Yacht Club was brilliantly illuminated and the
bridge and wharves were full of people and the city deserted. The
night was perfect and the scene was *brilliant, magnificent, granderous*. I never witnessed such a sight before and never expect to
again."

One suspects that "granderous" is an adjective of his own inven-
tion, born with "grandeur" in mind, but the point is that he *was*
moved by the excitement of the night, so much so that even in retrospect, it evoked the unfamiliar superlatives.

If there is less of a record of this kind of thing than one might
desire in order to know the man better, put it down to the dictates
of the age and especially of the Quakers. Still, there is a piece of
paper, yellow and brittle with age, among documents related to Matthew and his family and on it a verse, penned precisely in Matthew's
hand, and unattributed. Being as methodical as he was, it is likely

that Matthew would have known who wrote it—if it was someone other than he—or where he read it, if he copied it, and that he would have recorded the author or source on the paper. If he wrote it, being characteristically reticent—and mindful that nontheological verse was not encouraged by the Friends—he might well have chosen not to sign it.

On the paper is written, in part, "The moonlight silvered every bush and tree,/ Time in its noiseless flight was lost to me./ Yet as the rarest flower must fade and die,/ So must the happiest hours cease to be."

Whether it was original or copied, Matthew could have thrown it away at any time and since he was orderly above all, would have, had he considered it of little moment. And he did not throw it away.

Here is how Ricketson found this man of "well-placed benevolence and hospitality: Struggling through life with a physical affliction that might well have excused him from many of the more active occupations of men, his heroic fortitude ever sustained him and kept him busily engaged . . .

"The gentler nature of his mother, combined with the sterner one of his father, while he abated no fidelity to his strong and somewhat exclusive religious convictions, rendered him more congenial and accessible to one like myself, with whom he could not always agree . . ."

Matthew possessed an "earnest determination to succeed," and "married in early life to one who had already devoted herself to the cause of Christ, he found that support and encouragement to his own convictions that contributed largely to the formation of a character which, in its purposes, nothing could move or swerve from his path of duty. *Suaviter in modo fortiter in re* . . . (Calm in manner, but firm in purpose) . . ."

Furthermore, as a student, one of his best subjects was mathematics, which equipped him well for the countinghouse.

It is still possible a century later to walk the several routes that Matthew might well have taken from home to countinghouse each working day. There is, in addition, extensive reliable information available as to what and whom he would have seen, depending on the season. Consider then, the nature of his walk to work, his city, and his day.

From Hawthorn Street where Matthew lived, New Bedford slopes easterly to the river. On this day of his time, the sidewalks are of gray slate slabs. One must mind the footing, for the slates are slippery in the wet of spring and the fogs of fall.

Horse chestnut trees overhang the walkways and boys throw stones at the nuts to bring them down, shiny mahogany in their pale, thorny cases. (What good are horse chestnuts? To carry in a sack and look at, to trade, to carve into minuscule baskets and to feel, curved and smooth, in the pocket.) There are three or four magnolia trees on the route, especially at Loum Snow's and Thomas Hathaway's, beautiful in the blooming. Perhaps better than beautiful, they are symbols of well-being and prosperity, hallmarks of a civilized respect for the attractiveness of nature, colorful notations reminding that one's neighbors take good care of their property. (The "disorderly class," of course, did not respect property. Matthew's own summer place, "The Cottage," had recently been broken into. Seth Anthony's peach and cherry trees were constantly looted. Garden thieves in the melon patches were a persistent annoyance and many people had taken to letting out their dogs of nights as a consequence.)

Here is a black iron horsehead hitching post, with ring in the nose, at the sidewalk's edge, and a good piece of property on Third Street for sale, with grapevines, currant bushes, pear, peach and quince trees, rosebushes, gas fixtures complete, and pipe laid through the cellar for the Acushnet water.

There is City Hall. The motto of the city is *Lucem Diffundo* (We Diffuse Light) and what was meant originally, of course, was the light of sperm candles and whale-oil lamps. It is sobering to remember that twenty years ago, in 1851, forty-eight whaleships were added to the port's fleet in a twelvemonth and, at that time, if the whaleships of New Bedford and Fairhaven had been extended in line, they would have stretched ten miles and their whaleboats another six, with ten thousand sailors to man them. Times change.

Here is the Parker House, which has just installed something called Drake's Solar Gas Generator, its fuel supplied by a tank of gasoline buried in the backyard which supplies, one is informed, brighter light than coal gas.

There are fat oysters for sale in Bartlett's Market. A stuffed osprey sits in the window of Lowden's Millinery with a wingspread of more than six feet and a card beneath reading, "Shot on the Baptist Church at Nantucket." There is a sign: "Public bathrooms in the Ricketson Block will be open from April 1st every day and evening until further notice. Ladies accommodated from 12 noon to 6 P.M. each day. Private accommodations for ladies by applying to Mrs. Hines, the matron."

Here is an ice wagon of Hatch and Company, drawn by a well-fed horse. Hatch has done well with his seasonal crops—cranberries in the fall and ice off the same meadow in winter. Nothing like having your dish right side up when it rains.

Matthew passes Liberty Hall: "One Night Only, the Great Original Lettie Davis, female, a variety combination, featuring twenty brilliant artistes, from the principal New York and Boston theaters, together with a full orchestra brass band. Also, Flashes of City Life of Providence in 3 glimpses, with new scenery and stage effects; the Great Fire Scene and the Gambling Hell." The drop curtain in Liberty Hall represents the temple of Vesta at Tivoli, a favorite subject with painters of drop curtains. There is an unbelievable plateau in the picture on which are perched several people; there is no visible way up or down for them, and this has been remarked upon by audiences over several years. Israel Parsons is the manager of Liberty Hall; Joseph Wing takes tickets at one of the doors. One evening, Wing stepped into a side aisle to watch a performance and led in the applause for a singer who was called back for an encore. Mr. Parsons called Wing to one side and said, "Mr. Wing, I wish you would not applaud. It prolongs the performance and consumes the gas."

There is a fifteen-pound green crookneck squash in the New Bedford *Standard* window that Warren Ashley grew. There are bananas and melons in Rice's store, and Sanders Clothing has booklets on how to operate the new fire alarm telegraph. Daniel Murphy, carrying a bundle of leather, is heading for his shoemaker's shop on Ray Street; the man in the high hat is the German cigar-maker who works for Charles A. W. Oesting, and the woman is Miss Emma C. White, who retired as a teacher at the high school for reasons of health. They presented her with a bust of Dickens.

At the riverfront, the Fall River schooner *Naiad* is loading with ice at Ashley and Terry's. Smack boats with barrels of scup from the traps off Menemsha Bight are taking out the catch, which is bound for the New York market. Bark *Catalpa* has just arrived. Bark *Abraham Barker* is outbound from Commercial Wharf for the Pacific. Bark *Trident,* drawing too much water to lie at the wharf, is loading from a lighter at anchor in the lower harbor and, by the looks, nearly ready for sea.

And so, Matthew arrives at the countinghouse next to the Howland wharf, in whose southside berth the new *Concordia* had lain nearly four years ago, being rigged and outfitted for her first voyage. A good many had come to see her, broad stern and carved eagle turned to the stream and the figurehead on her graceful bow visible from the foot of North Street. *Concordia*, the span of oceans gone under her keel, the blood of whales across her deck, would soon arrive at Honolulu to prepare for 1871's season of Arctic bowheading.

The Howland countinghouse—any countinghouse of the period, for that matter—was less a place than an atmosphere. The ceilings were high and clean. Much of the furniture was heavy and dark and featured, more often than not, a good deal of bright varnish on tongue-and-grooved pine. To the degree to which paper and ink smell, and they do, the countinghouse smelled principally of paper and ink; the interior was quiet, undecorated, unimaginative, methodically stuffed with files, folders, portfolios, and ledgers. It possessed a Spartan aura; its inhabitants were serious. You had to go no farther than the worn threshold to stifle nonsense and time-wasting, even if the proprietors had allowed any. The full, but well-trimmed beards so often found there (Matthew wore one) contributed to the sense of inflexible mirthlessness, although, to be fair, in the period of whaling with which we are concerned here, there was little or nothing to encourage more than an occasional wan smile anyway.

What went on in the countinghouse, especially as the nineteenth century was winding down, was a matter of watching expenses in all aspects in order to stay afloat financially. In an effort to make money, the desperate contradiction not really faced was that, while extreme caution prevailed at home (the chief clerk in the countinghouse was instructed to open the morning mail by slitting the envelopes carefully so that the accountants could use them as scratch paper to figure on), the very fact of sending ships to the Arctic was expensive gambling of the headiest kind, done as if the men who owned them had money to throw away.

Watching the market's maneuverings and especially deciding when to sell was critical. ("Sales were made last week in New York of several small parcels of Arctic bone, 2,000 to 3,000 pounds at 70 cents gold per pound. The importation of whalebone [from the American Arctic fleet] has been made earlier than usual this season and the stock now on hand is larger than the average consumption, both at home and abroad, for the past eight years.")

There were so many areas in which attention to business made the difference between profit and loss. ("Notice to captains of the fleet: Act with the best interest of all parties in mind; keep a sharp watch for whales; when shipping oil home, if possible, make the merchant carrier liable for leakage, and make sure not to have too many natives aboard at one time . . .")

Innovation was suspect; what had worked ought to work always, and change was risky. Thus, the gear for whaling, from the harpoon to the clumsy, massive-barreled windlass that was formidable to repair because of the manner in which it was put together, persisted with little change for generations, symbols of a largely static technology. Francis Rotch, who tried unsuccessfully to apply technological advances to whaling, concluded, "The bare idea of novelty likely to produce change [in the New Bedford whaling industry] not only received no encouragement but was practically condemned without knowledge."

It was no superficial observation when Ricketson remarked of Matthew, "Conservative by nature, he held fast to that which had, to his mind, proved true and sound in doctrine and practice." What old George Howland had done successfully, Matthew—and his brother, George, Jr., as well—continued to do in essentially the same manner, even beyond that time when it became increasingly obvious that it was not working as well, or perhaps not even well at all.

Selecting good masters, mates and boatsteerers was of obvious importance if you hoped for a "greasy" voyage. Owner-agents pooled their knowledge of men and officers in a dossier that was circulated among the countinghouses when the companies recruited personnel for their ships bound out. These are some of the characteristic comments listed after names in the dossier: fair officer; dirty, not saving; very fishy (which means a good whaleman); good, but troublesome; good, but unlucky; no good around ice; A-1 master, but expensive; skunk; no good; crazy Frenchman.

The nature of matters with which Matthew Howland was necessarily concerned in the course of his working day are suggested by the following excerpts from letters to his son, Richard, the Howland agent in San Francisco, written between 1878 and 1881:

"In regard to the boats [whaleboats required to replace those lost or damaged by Howland ships operating in the Pacific] James [Allen] has been in and says he can forward two boats next week; one of thirty feet and one of twenty-eight feet; or if thee could wait a week or ten days longer, he could send two boats of thirty feet each. Will thee ascertain when a ship is to sail for San Francisco and write by return of mail and say what James shall do? I saw Wing and he promised we should have first chance to purchase [walrus] ivory on board *Syria*. Have not heard word yet from Wood about *Rousseau*. A letter from Hunt this morn says he has sold 1,000 pounds South Sea at $1.75, which is a *little* better than $1.40."

"I fell in with James Allen and told him the necessity of having the two boats go to Panama, ready immediately. He is going to hire help and work night and day until he has them ready, but I fear we shall not find a vessel to take them. There are so few going nowadays. Cleveland thinks there will be one leaving in the course of next week and I have charged him not to fail in letting me know. I suppose, as a last resort, those to go by Panama might go by propeller [steamer] to New York, but this would increase the expense. I believe the propeller charges $10 a boat; sailing vessels charge $6. I sent Hunt and Company bundles of bone by propeller, as a sample of ours. My price is $2 at present, but doubt whether I shall obtain it. I find it rather lonely in the counting room as George seldom comes in. Still, I have plenty to do, which helps out."

"James Allen was in this morning and said the papers reported that the Panama Railroad broke down so that one steamer had brought back her freight and reported the Road would not be clear before February. The boats were shipped yesterday, so had thee not better tell the Pacific Mail steamer folks to inform Captain Perry of Schooner *H. A. Taber* when he arrives with the boats [at New York] to send them to the ship at East River and deliver them to the next packet to sail for San Francisco? Please attend to this and see if the report is true; if so, have the boats delivered to the ship in the East River and thereby save expense."

"I called on Bartlett and he told me that his ivory was on board *Victor*, bound to New York, and not on board the *Syria*—and that he *would not* sell his ivory to arrive. He says also that he has ordered his captains not to kill walrus another season so as not to deprive the Indians [Eskimos] of what they depend on for their food; and, therefore, he thinks there will not be much walrus another season and prices will be higher. He did not say how much ivory was on board *Victor* and I did not ask him."

"Hunt wrote me yesterday, saying he could sell 1,000 to 2,000 pounds of our bone at $1.90, requesting me to write him today if he should sell or not. Well, when I was going from the bank to Meeting, I went over to the telegraph office (I don't know what made me do it) and told them that if any telegram came for me, to send it to the house. Well, it so happened that we had a long Meeting and when I arrived home, it was one o'clock and there was a telegram from Hunt, sure enough; saying he had sold 1,000 pounds at $2, to be sent tonight by propeller. Well, the propeller leaves at two, so I had to stir around, I can tell thee.

"So I ate my dinner, had the horse put to wagon . . . It was nearly twenty minutes of two when we left home. I found a teamer, helped him load, and had the bone (15 bundles) at the propeller a few minutes after two, *all right*. Pretty smart for old folks, I think. And now we have sold at $2, I think we shall be able to dispose of the remainder at the same price, which will be very satisfactory—if we can do it."

It was principally Matthew (because George, preoccupied with affairs of community, "seldom came in" to the countinghouse) who corresponded with the masters of the far-flung Howland fleet, and with all others who had business with the firm of "G. and M. Howland." Matthew's letter book, containing copies of this correspondence, is a revealing reflection of the man, his works, and his thoughts.

To a young Canadian who wanted to ship on a Howland vessel, he wrote: "The crew of a whaler are paid by a share in the oil taken; the share of a green hand would not be likely to amount to more than $350 or $400 at the end of a three-year voyage. [Scrutiny of comparable records suggests this figure does not take into account money invariably owed by the seaman to the ship for clothing and other supplies from the slop chest. A high markup was traditionally charged for such material—in the post–Civil War Wing fleet out of New Bedford, masters were instructed to charge "exactly double the wholesale cost"—and this item could very easily reduce Matthew's figure by 25 percent or more.] The promotion of an energetic young man, however, is rapid, since a large proportion of the foremast hands are ignorant blacks and men of mixed blood who have no ambition to rise."

When a young officer of the Howland fleet indicated that he wanted to come home when his three-year contract was fulfilled, in part, at least, because the captain was stingy with food, Matthew acknowledged that the officer had signed on for three years and could, therefore, leave the ship at the end of that time if he chose. Matthew nevertheless went on to write, "We know that was the expectation *then* but it looks now as though it would be necessary to lengthen the voyage a little in order to make it a saving one and consequently we hope, as the three years will be up in July, right in the middle of the [whaling] season, you will consent to remain on board that season out or until October next, if we *desire* it and we do most *earnestly*.

"As to Captain Almy's recruiting ship [supplying provisions] so sparingly, we will write him and request him to be more liberal in future."

What is more, Matthew *did* write to the vessel's master on that same date, tactfully, yet pointedly, as befits a successful man of business. He noted, "Since writing thee 3d inst, I have heard reports floating round that you did not give your men as much fresh provisions as the other crews had. Now, of course, we do not know as to the truth of these reports but if you should find, on looking back, there had been a little shortness in this respect, we would encourage you to deal fairly with your men in regard to their food, as we all know that if Jack is well fed, he does his work more readily and cheerfully, there is no grumbling, and everything goes on more smoothly.

" 'A word to the wise is sufficient.'

"P.S. If the [supply] schooner should not arrive before you are ready to sail for the South, would it not do to cruise around the island a week so as to be sure and take the provisions on board this fall and avoid expense?"

And there were other problems involving human relations. To Captain F. W. Vincent, Matthew wrote, "How fast time goes; you are already thirteen months out. Did we not have some such understanding as this when you shipped, that if you did not get oil so as to be making a good voyage, your wife was to come home? I hope you will get oil so she can remain with you. With kind regards to Mrs. Vincent."

It was Matthew, precise and practical, who devoted an entire page of his careful penmanship (listing all the vessels' names) to calculations concluding that, in a ten-year period, a yearly average of 1.7 vessels engaged in sperm whaling, or 1.5 percent of the New Bedford fleet, had been lost at sea. Although the page is without comment, one assumes he concluded this was a reasonable risk and an acceptable loss.

To another of his shipmasters, he wrote, "We must say that we, too, were disappointed in your catch of sperm oil last season out, but we ought not to complain when you say that you are satisfied you took all you had a chance to take . . ." But Matthew could not leave it at that, even though he had begun graciously. For he *did* complain, adding, "Where was the *Ocean* cruising, that she should take 700 barrels sperm oil while you were taking 700 barrels humpback? She stole a march on you, somehow or other . . ."

To one captain, he observed, "I thought I would write a few lines to say that we received thy letter dated July 27th at Honolulu, reporting the Kodiak grounds a failure as far as you and some other ships were concerned, though it was not to some ships, as there were quite a number took from 600 barrels to 1,500 barrels of oil . . .

"I am in hopes you will get a good *cut* at New Zealand and be reported into the Islands [probably Hawaii] in the spring with 800 *barrels*.

"I have no doubt I said quite enough in my last letter about expenses, but do try and make them as low as possible as that is our only hope of making money this voyage, considering the present price of oil and bone.

"Edward S. Taber unites with me in kind regards to thyself, also to mate and 2d mate, whom I presume are good officers and whalemen. Who is 3d mate now, and are all the boatsteerers good?

"Hoping and believing the next accounts will be more favorable, I remain, very truly and respectfully . . ."

With Captain Robert Jones as master, *Concordia* sailed from New Bedford December 7, 1867, for the North Pacific. On March 3, 1871, Matthew wrote to Captain Jones, who, presumably, was then in Honolulu, preparing for a season of bowheading in the Arctic:

"Respected friend—

"We think, on the whole, the *Concordia* had better come home after this season.

"I am in hopes your drafts next 'fall' [Matthew always enclosed "fall" in quotation marks] will not be as heavy as last 'fall' although I suppose if you get a good season's catch, it will take considerable money to pay off the crew.

"I wish you would do what you can to induce Captain [Valentine] Lewis to go the third season in the *Thomas Dickason*. [In the course of events, George, Jr., and Matthew had disposed of the *George* and *Golconda*, while the *Corinthian*, commanded by Captain Lewis, was lost on Blossom Shoals in the Arctic whaling season of 1868, and had replaced them with the *Desdemona*, *Clara Bell*, and *Thomas Dickason*, thus maintaining a ten-vessel fleet.] Tell him you think he will make a mistake to leave. Don't let him know I have written you about it. He wrote me last fall that he may conclude to leave after this season and come home. Do you think he dislikes the *T. D.* or what is the cause? [Matthew's sharing of confidences with Captain Jones was unusual; this, coupled with the fact that Jones was commanding the Howlands' prized $100,000 vessel and that there is no real criticism of his management in this letter suggests he was exceptional and that they thought highly of him.]

"Your first season's catch sold for $31,000; last season, $52,000, making $83,000. Expenses to come out—I have paid $31,500 in currency for drafts; freight of oil and bone, $6,500; provisions etc. from *Corinthian*, and casks etc. sent out, $6,500, and about $40,000 advance to officers, and sundry expenses, so that the balance now to credit of ship is about $28,000. About $40,000 oil and bone now on the way, making $68,000, so that if you should be so fortunate as to take a good catch this season [in the Arctic], we may get a new dollar for an old one and perhaps *a little more*.

"I have no complaint to make, Captain Jones, in reference to your drafts as I have no doubt you have been as saving as possible.

"What a severe time you must have had in the Arctic last season. *Concordia* must be a splendid sea boat. I don't see how you worked the ship with all that ice on the bow. However, I am in hopes you will have a better time for the season now coming.

"Now if nothing happens to you beyond what is usual, we may expect to see you again in about ten months.

"Brother George writes with me in kind regards to you and hope and trust your success will continue."

The warmth of Matthew's letter to Captain Jones may have been a reflection of the fact that, in the more than three years since his departure, Captain Jones had shipped back to New Bedford 164 barrels of sperm oil; 3,503 barrels of whale oil, and 39,965 pounds of bone. Jones had been a captain of Howland vessels for at least a decade, having previously been master of the *George and Susan*, in 1857, and of the *George Howland*, in 1862.

So the day went. Finally it was time for Matthew to go up the hill to home again, walking into the late sun—walking, at least in the years before "age and disease had begun their work," with vigor and dispatch.

Matthew had a reputation for being a good host, warm and generous, and if there was no Lyceum lecture, no meeting in behalf of God or country, then there would be an evening at home, very likely with guests. For dinner, perhaps a braised saddle of lamb with a border of baked onions and potatoes; for dessert, a boiled pudding and sauce, or plump native blueberries, depending. Maybe some fresh vegetables from the garden at The Cottage. And afterward, an hour or two of sensible talk.

He wrote, on one day, "The weather is very pleasant and mild. Yesterday, there was just enough snow and ice to make good sleighing and I had a nice sleighride with Tom. But now it is thawing fast. J. Grinnell and wife and C. Coffin and wife are coming to dine with us and will be here soon . . ."

It was, indeed, "thawing fast," economically as well as otherwise, and the wonder of it is that this man, so sensitive, so astute, so shrewd, did not acknowledge (or was reluctant to acknowledge) that trouble was brewing in whaling—especially in Arctic whaling—and that the nature of it was common talk among shipmasters and that it had been well documented almost twenty years before.

# 8.

# Letters from the Arctic

The master [of the whaleship] was in charge of life and property and his word was law and where he willed, he could go. On his discretion and good judgment turned success or failure for all. His draft in foreign ports for supplies or requirements bound every individual owner in the ship for the full amount of his disbursements. In this respect, the power entrusted to him illustrates the inconsistencies of human nature; close, careful men who, on shore, would not trust their neighbor with a small portion of their property, who distrusted everyone's judgment and integrity, would placidly repose in the power of a master who was to sail around the world and had the right to make drafts in any quarter that might easily absorb their all.

—Hutt's *History of Bristol County*

During the final two months of 1852, the *Whalemen's Shipping List and Merchants' Transcript* published a series of unsigned letters about the prospects and perils of Arctic whaling. Although the author of these letters is not known or certain, it is likely to have been either Captain Asa Tobey of the *Lagoda* or Captain Charles A. Bonney of the *Metacom*.

In 1852, George Howland, Jr., was forty-six; Matthew Howland, thirty-eight, and the firm of G. and M. Howland was some three decades old.

The letters offer an unusual insight into the thinking of a whaleship master, including some reactions to what he was doing that were not normally shared even with the owner of the vessel of which he was in command.

*At Sea, November 22, 1852*

The great failure of the last season was occasioned by circumstances over which man had no control. I followed along the verge of the ice from Cape Thaddeus to St. Lorenzo's Island, and so on along the eastern shore of the Straits, until off Cape Prince of Whales —a distance of about six hundred miles. We saw but few whales, and these, as soon as the harpoon pierced their sides, would immediately sink below the surface, nor rise again until they had reached the icepack.

Often, I have looked out upon the field of ice in clear weather, anxiously watching for clear water, and watching in vain. It was indeed a dark and a hard season.*

I thank God, in the great and unparalleled destruction of property, that so many human lives were spared. I felt as I gazed upon the great frozen icefields stretching far down to the horizon that they were barriers placed there by Him to rebuke our anxious and overweening pursuit of wealth.

Perhaps this is as good a place as any to speak of that peculiar animal, the polar whale. This differs in form and movement materially from the right whale, although it resembles the latter more nearly than it does any other species. In motion, the polar whale is not unlike the sperm. Its adaptation to the frozen region which it inhabits is very remarkable. The thickness of the blubber, which is a great retainer of heat, enables it to remain in the coldest water without in the least checking its powers or faculties. In fact, this whale prefers water nearly at the freezing point. Some of them exhibit great anxiety and haste to move north, although others linger further south during the season. The peculiar form of the head is admirably adapted for moving among ice, which it would, from want of breath, be absolutely impossible for any other whale to do, since where the ice is close-packed and heavy, it would be impossible for them to

---

* The writer referred to the fact that, in the bowheading season of 1851, the following whaleships were lost: in June, the *New Bedford*, on Fox Islands, with the loss of four of her crew; in July, the *Henry Thompson*, in the ice, near Diomede Island; the *Houqa*, near Cape Oliver; the *Armata*, on a reef near Cape North; the *Superior*, in the Anadir Sea; and in August, the *Globe*, on East Cape, Bering Strait.

raise themselves up to the surface, and as much so to get their spout holes above water, in the small cavities and blowholes in the ice, on account of the head, while the long bow head and high spout holes of the polar whale enable it to rise to the surface and spout with ease, where a right whale could not find the air.

Still, I think the polar whale finds great difficulty sometimes in raising its spout holes above the water in these small cavities, and where the ice is very thick.* This difficulty is guarded against, however, for it has the power of retaining its breath for a great period of time. They are many times much exhausted in passing under extensive fields of ice.

On the 29th of June, 1851, I had reached King's Island by passing to the south and east of St. Lawrence Island, and had followed the ice along on about a north course, from one island to the other. At 4 o'clock P.M., I discovered a great number of spouts to the northwest, over two long points of ice extending a considerable distance from the main body, and about eight miles distance, in a bay of clear water formed by ice. In fact, for about a mile in extent, the air was constantly full of spouts. They remained in that position as long as it was sufficiently light to distinguish them. From the unusually large and high spouting, it was evident that the whales were greatly exhausted from having come so far under the close-packed ice.

They entered the ice from seventy miles to the east of Cape Thaddeus to St. Lawrence's Island, and must have gone from one hundred to two hundred and fifty miles under the ice. I am confident that there were no lagoons or openings in the ice with the exception of some small cavities, or blowholes. The wind had been blowing from the south and east during June. This, with a northeast current, must have wedged the ice as close as possible, and no clear water was seen by ships that cruised along the verge of the ice, nor by those that worked their way some distance into it. H.M.S. *Enterprise* entered the ice near Cape Thaddeus, crossed the Anadir Sea, through the Straits, and 140 miles north of the Diomedes without seeing any clear water with the exception of a narrow strip, on the west shore, north of Cape Chaplin (?).

---

* Captain Charles Brower has said that a bowhead will break ice two feet thick.

Captain Colinson told me that he measured ice eighteen feet thick and in some places it was so heavy, and so closely packed, that he could scarcely make his way through it with his ship and he was nearly a month in sailing that distance.

The whales that passed up by the northwest cape of St. Lawrence's Island were going a direct course for King's Island. Their instinct must have taught them that there was clear water there, and along that coast, and as I observed their course, I was led to believe that there must be clear water in that vicinity and that I should find whales there, as I did. They had only stopped a short time to rest on their way into the Arctic after a tiresome passage under the ice.

No other but the polar whale could have possibly made the passage under the ice, for such a distance, and it must have been difficult even for that species. The breathing places are a hole in the ice, or where two irregularly shaped cakes have left a small aperture about as large over as the breadth of a whale's back. This would be about one third his length and extending as the ice does some ten or fifteen feet below the surface of the water, a whale, to raise his spout holes to the air, must almost double himself. I do not think, as some do, that they prefer the ice to clear water, although they are frequently found amongst large floes of open ice, appearing perfectly at home.

I think this is owing to the abundant food which may be found where the ice floe is very large. When in the ice, the whale is very still, and moves easily. When I worked up towards the whales off King's Island I saw over the points of ice, it was midnight when I reached the spot and but three whales were seen. These were going quickly to the north. I sent two boats in pursuit, and two into the ice, which was open, so that they could work their way in for a mile, but not another whale was to be seen.

This confirmed me in my opinion that these whales stopped to rest when they got into clear water, but how long, I am not able to determine.

The ships *Hobomok* and *Ann* passed King's Island one day in advance of me, saw a good many polar whales north of the Diomedes, going north fast along the verge of the ice, and from one point to another, they followed on to Point Hope. Their course was obstructed by the ice closing with the land, but the whales still kept on, affording another proof that clear and open sea did exist to the north of this cape, notwithstanding the straits were nearly full, and the Anadir Sea at the time . . .

*At Sea, November 25, 1852*

I closed my last letter with some discussion on the subject of clear and open seas north of Cape Hope. It seems to me very certain that whales would not pursue their way under and along extensive fields of ice unless there was clear and open space beyond. The whales I spoke of seeing in my last were an early school, hurrying towards the first or earliest food, which, in my opinion, consists of large shoals of small fishes. This is the first food of the whales in coming from the south—next is the shrimp, and minute insects with which the water is literally filled.

If I wished to give an idea of this last-mentioned provision for the whale, I should say that you would get a good notion of its appearance by throwing a handful of pearl barley into the water. These little mites are very fat, and where they exist in great numbers, the water is nearly covered with "slicks" caused by the oil which rises from them. These, and the shrimp, too, I think, retain nearly the same position during the year.

After they lay their eggs, they sink to the bottom, and there remain until the next season. When the ice clears and the water gradually becomes warmer, they soon come to maturity and gradually rise to the surface. Should the ice remain late, or entirely cover the surface, it is my opinion that these two kinds of whale food never come to maturity—and this is the reason why whales were not seen in the Straits after the ice cleared. Only a few were seen and they moved on through at a rapid rate, only excepting a small number between King's Island and the Diomedes.

It must be remembered that a passage was open during the season on that coast, and the water one mile from the ice is from two to five degrees warmer than that in its immediate vicinity. It is certain that the shrimp do come to maturity on the bottom. If you take a piece of lean flesh from a whale, and sink it to the bottom, and let it remain there a few hours, it will be covered with a large white worm. A shipmaster who lay at anchor for some time about Cape East the season previous told me that he watched their growth in this way. When they were from two to three inches in length, they began to change color from white to red.

The whales remain at the North until the water begins to freeze over, then work gradually south, and frequent the bays and shores—following up these small fish as they move south. This is the time, in October, when the natives lay in their winter supply of blubber. "When the small ice comes," they say, "then plenty of whales." As the season advances, they are forced to move south both on account of ice and scarcity of feed. For the support of the whale, these waters are very rich and productive, and regularly produce and bring to maturity immense quantities of living "mites" and small fishes—as regularly as the rich soil does an abundant crop. But as this last may be cut off from a variety of causes, so the crop of "whale feed" in the Northern Seas is sometimes diminished and sometimes entirely destroyed. This may be occasioned by the ice remaining very late and entirely covering the waters beneath which the germs exist. This undoubtedly causes the difference in the movements of the polar whale, their different route, and positions in feeding at different seasons.

During the two seasons in which I cruised in the Arctic, with a few exceptions, the movements of the polar whale have been entirely different. In the localities where great numbers were found the last season, they were scarce during the first, and vice versa. This is one reason why some ships did not take more oil. The instinct of the whale teaches him where the best "feed" is to be found and he goes there. Nor is the polar whale the sole consumer. There is a small gray whale called the "California Gray" and by others the "Muzzle-Digger" and by others still, the "Scamperdown." In the Anadir Sea are humpback whales and some finbacks. The walrus and seal are also very abundant, and birds of various kinds are so numerous as literally to cover the water and fill the air in every direction.

There are seven or eight different kinds of ducks, the most numerous resembling the "old squaw" of our shores. These birds move northward about the same time with the polar whale, some going far north into the Arctic, and others remaining in the Straits and Anadir Sea. Their food is the same with that of the whale.

There is a small whalebird that very much resembles a Mother Cary's chicken in size, form and motion, of a gray and reddish brown, with broad stripes, that is very common wherever the whale "feed" is abundant. They feed on the oil that rises from the minute animals which I have described. Their beaks are in constant and rapid motion when setting on the water amidst the oily slicks. These birds move with the whale and sometimes in advance of him, but very seldom get further north than 71 degrees and there is seldom much clear water north of that.

The *Enterprise* went as far north as 73 degrees and 20 minutes, but saw no polar whales during her cruise. I think it probable that they may pass round to Davis Straits in some favorable seasons, when there is considerable clear water, but I believe some of them go as far south in the winter season as the Japan Islands, Matsmai and the Kourile Islands, occupying as many degrees of latitude in the winter as in the summer. It is positively known that they occupy 15 degrees of latitude in the summer, for they have been found thus far apart, although they were not numerous at either extremity.

When they move, those that are furthest south go on still further, and generally keep in advance. Very few remain north of 52 degrees in the winter. They have been seen south of Bering's Island the latter part of May, and very few have been much north of that island earlier than the 29th of April. They calve at all seasons of the year. Calves are not seen in the Arctic, because the whales go into the bays and bights to produce their young. Very few go into the Arctic to calve. They remain along the coast, southwest of the straits and Kamchatka.

Captain Sayer of the ship *Mary Mitchell* saw a calf in the ice in

June, 1851, in the Anadir Sea. I myself took one out of a cow September 12, 1852, in the Arctic that measured only six feet, three inches, and would not certainly have been born before December or January. This shows that the time of calving varies about six months.

I heard it from a Russian officer that he had seen the beach covered with dead calves in the winter season, about the entrance of the Bay of Petropaulovski. Last season, the ships in the Okhotsk Sea found a good many calves in the bays at the southwestern part of the sea and I have been informed that some of the ships made a business of taking calves that yielded from ten to twenty-five barrels of oil, with but little, if any, bone.

I have heard much said about the mysterious movements of the Polar Whales, their sudden appearance, and equally sudden disappearance, and I know some have believed that they could remain underwater for any length of time. This is not strange when you may cruise so long with a good lookout and not discover a single spout, for a distance of three or four miles and then, in an instant almost, discover whales in almost every direction and even directly astern of the ship.

After some thought, I attribute this to sudden change in the atmosphere and also the different ways in which the whale spouts. When he is tired or frightened, he makes a much larger, thicker, or higher spout than when at ease. The atmosphere is, at times, very dense along the horizon, and the surface of the water of so light a cast that the spout of a whale can be seen only at a short distance even when it is strong. A whale is often discovered by the spout holes when no spout is visible. This is when he is very still, having had his fill, so that he does not exert himself below the surface, and when he rises, breaks very easily, and keeps his spout holes underwater, heaving up very little with their breath. For instance, a ship may be standing along, the weather good, and the seeing apparently so. One would suppose a whale could be seen two or three miles, if there were any within that distance, but still there are not even the slightest indications of his vicinity so quiet is he, and so thick is the horizon.

Besides they are often stretched along almost in a straight line and, remaining underwater, sometimes a ship might pass along and arrive in their midst before seeing them. The vapor or smoke then rises from the horizon, the surface of the water becomes dark, the whales again commence feeding, and one after another, the spouts rise in all directions, stronger and stronger. This sudden change in the atmosphere and in the manner of spouting accounts well enough for the sudden appearance of whales, although I think they can remain under from one to two hours.

*At Sea, December 15, 1852*

Perhaps a few suggestions will not be out of place here relative to the future fortunes of Arctic whaling. In this, an immense amount of capital is invested. It is, therefore, of vital importance that everything possible should be done to sustain it—for without the most watchful vigilance, it must ultimately fail.

In the commencement of right whaling, the Brazil Banks was the only place of note to which ships were sent. Then came Tristan, East Cape, Falkland Islands, and Patagonia. These places encompassed the entire South Atlantic. Full cargoes were sometimes obtained in an incredibly short space of time—whales were seen in great numbers—large gams and bodies of whales were often seen where they had been gamboling unmolested for hundreds of years. The harpoon and the lance soon made awful havoc with many of them and scattered the remainder over the oceans and many, I believe, retreated farther south—a few remain, as wild as the hunted deer.

Can anyone believe that there will ever again exist the same number of whales? Or that they multiply as fast as they are destroyed? I have seen in print the statement of some wise person who did not believe that it was possible for the fleet to diminish the number. Had this been the case, at the end of six thousand years, it appears to me the entire surface of the ocean would have been literally covered with whales.

After the Southern Ocean whales were well cut up, the ships

penetrated the Indian and South Pacific Oceans, St. Paul's, Crozettes, Desolation, New Holland, New Zealand and Chile. I believe it is not more than twenty years since whaling began in either of these localities—but where now are the whales, at first found in great numbers? I think most whalemen will join in deciding that the better half have been killed and cut up in horse pieces years ago. A part of the remainder have fled further south. A few yet remain, and most of them know a whaleboat by sight or by sound. This completes the southern circumference of the Globe.

Then came great stories of large whales in large numbers in the North Pacific. The first voyages by their success created great excitement—the fleet there increased and was fitted out with extra care and skill, and in a few years, our ships swept entirely across the broad Pacific and along the Kamchatka shores. They moved round Japan and into that sea and there whales were found more numerous than ever. The leviathans were driven from the bosom of that sea, their few scattered remnants running in terror whenever the enemy is near.

Then the great combined fleet moved northward towards the Pole, and there the ships of almost all the whaling ports in the world are and have been for several seasons, lending their united efforts to the destruction of the whale—capturing even the young. These polar whales were most easily captured at first, but already it has become difficult to do so, for they are fast becoming shy.

But the general subject of Arctic whaling is too extensive to be discussed in this letter and I will resume it in my next.

*At Sea, December 22, 1852*

I spoke in my last of the fact that, while at first the polar whale was most easily captured, his nature had been entirely changed by constant and untiring pursuit. He is no longer the slow and sluggish beast we at first found him. Particularly at the latter part of the season, they are very shy. I have often noticed, after one or two whales were struck in the morning, after the fog cleared, that the entire body of whales would be stirred up so that it would be almost impossible to strike one during the whole day. Within a space of from ten to twelve miles, there would be from fifteen to thirty ships, all doing their best, but the greatest number were to be seen without any smoke.* I counted fifty-eight ships and only twelve of them were boiling and I have seen a much smaller proportion in smaller fleets.

I know that the whales have diminished since I was here two years ago and they are more difficult to strike. How can it be otherwise? Look at the immense fleet, stretching from Cape Thaddeus to the Straits! By day and night, the whale is chased and harassed—the fleet perpetually driving them, until they reach the highest navigable latitudes of the Arctic. The only rest they have is when the fogs are thick and the wind is high. There could not have been less than three thousand polar whales killed last season, yet the average of oil is only about half as great as it was two years ago.

This fact speaks for itself and shows that it will not long be profitable to send ships to the Arctic. If the ships were to withdraw for ten years, you might again have good whaling. Would it not be the wisest course to pursue? Would it not be better to draw off most of the fleet from the old whaling grounds and to turn its energies upon new ones, if they can be found?

I cannot help thinking that there must be an immense number of whales in the Antarctic region which have never been troubled by our hardy whalers. Some search has been made, but nothing has been done worthy of the importance of the enterprise. An expedition fitted out expressly for exploration in those seas would, I think "pay" in the long run. It must, however, be a competent and thorough one—one that will decide the question of the existence of whales in that quarter for all time. This is something in which all shipowners should take an interest.

An express expedition would not cost much and if the expense was divided among our owners, it would be found individually to be very light. The question is one which it is very important should

---

* From the tryworks on deck, that is, where the oil was boiled out of the blubber.

be settled. If whales are found in that southern region, the interest rendered for the small outlay would be immense. The expedition should, in my opinion, consist of four staunch ice ships built expressly for the purpose, upon the latest and most approved models for speed and strength—without stem or keel projecting—the rudder of unusual strength, so constructed that it can be unshipped and triced up across the stern in a very short time.

The bows should be from eight to ten feet thick, of solid timber— the sides from two to three feet thick, in that proportion throughout and bolted in every direction. This will add to the combined strength of the hull and render it capable of standing heavy pressure through the icefield in case of emergency. This would enable the ship to pass through the fields of ice and to explore the distant open seas. This alone can render the search for whales a thorough one. Very much would depend on the rig. This should be snug and staunch and of the most improved plans for easy working. The speed of the ship also is an important point. She must be able to work off and keep clear of the ice when an ordinary sailer would be caught by the mysterious movement of the ice.*

These ships, even if the enterprise failed, would be very useful for Arctic whaling. If there should be any difficulty in manning them for "lays," I should recommend a system of moderate wages and good lays, such as would induce the most enterprising and hardy officers and men to enlist in the service.

I think such ships could be built and fitted for two and a half or three years for sixty or seventy thousand dollars each. I merely throw out these hints—they seem to me to be of vital importance. Whether they will meet the views of your readers or not, I do not know, but I am certain if the business of right whaling is to be kept up, something of this kind must be done, either by a joint stock company, or by some other means. . . .

---

* *Concordia*, designed for speed, and with double topsails—each roughly half the size of its single, cumbersome predecessor—for easier handling, was concrete acknowledgment of this thesis.

*At Sea, December 25, 1852*

. . . we need an expedition to survey thoroughly the whole Pacific and to give us better and more reliable charts and this can only be undertaken and should be undertaken by the government.

After whalers have explored the whole Arctic, from Fox Islands to the Barrier, government sends a fleet to explore and survey those regions. I believe that there is hardly a dangerous point or spot there but what has been already discovered and noted. This has been accomplished not by the aid of charts, which are very inaccurate.

Whalers have fairly felt their way through the fog and darkness. The survey, however, will have its advantages if thoroughly performed. The difficulty will be in getting the latitude and longitude in thick fogs and rainy weather. I think, however, that a thorough survey of the Pacific is more needed. If this could be accomplished in connection with Lieutenant Maury's chart enterprise, the advantages would be immense. American ships would no longer sail by foreign charts, but by their own—a circumstance which I think would be of great honor and credit to the nation.

# 9.

# "Our anxious and overweening pursuit of wealth"

I felt as I gazed upon the great frozen icefields stretching far down to the horizon that
they were barriers placed there by Him to
rebuke our anxious and overweening pursuit of
wealth.
—New Bedford shipmaster, commenting on
Arctic whaling in the *Whalemen's Shipping
List and Merchants' Transcript*.

Among the rules of discipline and advices of the yearly meeting
of the Society of Friends in Philadelphia in 1894, is the warning to
members against "applying to those who, by any art or skill whatso-
ever, pretend to a knowledge of future events . . ."

So perhaps, even as the year 1871 advanced, the Quaker whaleship
owners of New Bedford would have scoffed at the possibility that
anything unusual was likely to take place in the Arctic that season,
although subsequent events revealed that the Eskimos had, in fact,
predicted something unusual.

Nor is there any evidence that either George or Matthew Howland wondered, as had the author of the letters in the *Whalemen's
Shipping List and Merchants' Transcript*, whether God had intended
man to pursue whales beyond the Arctic's icy barrier. Or whether
anything untoward might occur if he had not.

Yet both brothers must have read the letters, must have thought
about them, at least briefly, even if they did no more than dismiss this
particular reference. For the periodical in which those letters appeared was highly esteemed; it had been in existence since 1843; it
had a worldwide circulation and never had a competitor. It was the

bible of the countinghouse and of the industry and, as George and Matthew read the Bible of the meetinghouse regularly, so they must have followed the *Shipping List* religiously, simply because it was practical to do so.

Moreover, they were living in, even leading, a theocratic society —albeit a society on the decline—and assuredly would have found it reasonable that a series of articles on whaling would include speculation as to God's intentions concerning whaling. The respected publisher of the *Shipping List*, Henry Lindsey, who was totally unlikely to publish nonsense or romantic fancies for his very practical readership, also obviously found it reasonable. To appreciate how far removed we are from what then prevailed, imagine that the *Wall Street Journal*, or *Barron's Weekly*, in an article on the depletion of fur-bearing animals, raised the question as to whether God had placed some of them in the deepest jungle to give them sanctuary and that perhaps man ought not to use helicopters to go after them there.

Finally, George and Matthew undoubtedly knew—even better than we do now—who the author of the warning articles was. There were not that many whaling masters and, what is more, the author revealed his recent travelings in detail, which would have narrowed the field. Also, he was exceptionally competent, a thoughtful, articulate, and philosophically inclined captain, and there were not that many, which is, of course, why Lindsey invited him to write the pieces, which the author noted in the first of the series. Lastly, it is reasonable to assume that the author of the articles, on grounds of competence alone, would have commanded respect within the industry for that was exactly what *did* command respect. If he wondered whether God had intended an Arctic whale fishery, others would wonder, too.

Still, probably no one within the industry wondered what God might do about it, if anything, if He didn't like it. In part, this was because New Bedford generally and the Howlands particularly were involved in many other ongoing matters, some having to do with God and some not.

For example, the thirty-seventh anniversary of the New Bedford Bible Society, a group composed of members of the Trinitarian, County Street, Pleasant Street, Unitarian, and Middle Street Christian Societies, was held in the North Congregational Church, and was well-attended. After an anthem by the choir, Matthew Howland, president of the society, read a passage from the Book of John.

The first speaker was George Marston. He commented on the influence of the Bible on the civilization of our country and he said that the settlement at Plymouth, really the origin of our national principles, was preeminently an outgrowth of the Bible's influence. He added that the Pilgrims crossed the ocean in pursuit of religious freedom and were tolerant of all creeds because they had bitterly experienced the evils of intolerance.

Matthew's wife Rachel was the next speaker and she forcibly pleaded the brotherhood of man as making the spread of the gospel a duty, comparing the efforts made in this direction with the efforts of businessmen to acquire riches. She said, "The Bible says thou shalt not steal and if it had been obeyed in a sister city [presumably, Fall River], how different would be the situation of some who have disgraced themselves."

In what obviously was an oblique response to Mr. Marston, she added, "The circumstances of the present occasion indicate a great advance of tolerance and Christian unity since the days of the Pilgrims. The charity of today allows a member of the Society of Friends to occupy a Congregational pulpit and, in another manifestation, it must provide for the spread of the Bible. This is of more importance than the ornamentation of churches and if means cannot be provided for both, churches should be plainer."

In what certainly was a direct response to Rachel Howland, Dr. Quint, the North Congregational minister, then said that he had been requested to speak in relation to the collection but that he desired to say one thing more. He observed that the Pilgrims at Plymouth were never persecutors, that the Puritans at Massachusetts Bay passed some savage laws and were not particularly gentle in their enforcement and that the few cases of intolerance in the Old

Colony were the result of extension of the legal authority of Massachusetts Bay.

He said he was liberal enough to exchange pulpits with the Friends minister any day, but doubted the acceptance of his proposition. Dr. Quint concluded by saying, "The congregation has listened to the crystal argument and felicitous expression of one speaker, the music of another, the eloquence of another and are doubtless wishing they could speak as well. Any of you can make a better speech by a liberal gift to the cause. A covetous man cannot enter the kingdom of God."

While the collection was being taken, Matthew said that once at a business meeting of Friends, one had spoken of the necessity of faith in connection with the subject under consideration and another had responded, "Faith makes thin soup; money is what we want."

Obviously, even the passage of a couple of hundred years, give or take a few, had not made it any easier to set forth a totally acceptable account of what happened to whom at Plymouth in the seventeenth century.

And from New York, Rachel wrote a letter to the editor of the *Republican Standard* signed simply "Howland," which is interesting because it was the only signed letter published in more than a twelve-month period (it was customary to use pseudonyms) and, I suspect, after having read all of them, it was the only letter to the editor written by a woman.

She wrote: "The Brooklyn presbytery has just concluded the first act of what is already a celebrated farce. A grave charge was preferred against the Rev. Theodore I. Cuyler, one of the most earnest, zealous, God-fearing ministers of Christ who labor in this or any other land, for admitting a woman preacher into his pulpit.

"During the stay of several eminent ministers of the Friends Society, whose public labors in the Christian work here was crowned with great success, Mr. Cuyler cordially invited Sarah F. Smiley, one of their most eloquent and persuasive preachers, to speak from his pulpit. She accepted and every hearer was delighted with the sweet messages which flowed from the lips of this richly endowed Quakeress. In Great Britain, she had been honorably received; the best minds there had honored her with their presence; she had preached from Presbyterian pulpits in New Jersey, Poughkeepsie, and elsewhere, but no public notice was taken of it.

"Alas, for Dr. Cuyler, the LWKSCA (Let the Women Keep Silence in the Churches Association) went for him and brought him before the ecclesiastical court of justice to show cause for allowing a woman (a woman!) and she a modest, unassuming, highly cultured Quaker, to preach in a Presbyterian house of worship.

"They may stifle the fire for a while, but the smoke is seen issuing in various directions and, ere long, the flame will burst forth and woman's gentle cheering, encouraging voice be heard in the Presbyterian household.

"Let us sustain the course of Dr. Cuyler. I am ready to hold up both hands in favor, not only of encouraging woman in preaching in haunts of vice, where her good influence is wonderful, but in heralding goodwill to men (even) over the cushioned pulpits of every denomination. To this end, I subscribe my name in full, not ashamed of my position."

New Bedford's Howland Mission Chapel was opened and dedicated, and the *Standard* commented approvingly, "Mr. Howland, with a kind generosity which is and should be appreciated, has built and furnished the chapel and the resident ministers of this city have generously supplied the pulpit on Sunday evenings. We hope to see the chapel crowded every Sunday night hereafter . . ."

Meanwhile, civic matters were making their demands upon both George and Matthew. Citizens of New Bedford learned of the great Chicago fire in a page one newspaper story with a ten-part headline and it says something for the community that, within hours, a well-attended meeting was held at City Hall to determine what aid might be given to the fire sufferers. The wealth of the city was represented at that gathering by its most active businessmen and it was obvious from the opening moment that they were disposed to respond liberally to the appeals of the burned-out people of Chicago. Matthew was chosen secretary; a committee was selected to solicit contributions—by the end of the meeting, $5,000 had been pledged and there were indications of several thousands more to come.

As for George, he was chairman of a gathering "packed with the intelligence and patriotism, the good order and temperance of the city," called by the proposal of the ward and city committees to re-endorse the incumbent municipal administration by renominating Mayor George B. Richmond. Characteristically, George remarked, as he assumed the chair, that he had supposed those present might long since have wearied of seeing him in a public capacity and he further assumed the applause that had greeted his appointment was due exclusively to the overflowing enthusiasm of the occasion.

He said he saw no need to make any extended remarks because everybody knew why they were there and further knew as well as he did the record of the incumbent administration during the last two years. Mr. Richmond was thereupon renominated by acclamation, accompanied by what was described as great applause, and he was subsequently elected mayor of the city for the third time by a plurality of eighty votes.

It was, in short, a bright and bustling golden age for the city in so many categories as to overshadow what now are obvious storm clouds of social inequity and change. One wonders now whether Rachel, essentially more concerned with the needs of people, while the brothers responded to those of institutions, may have been more sensitive to what was happening in New Bedford.

Yet it was easy not to notice. The Potomska Mill buildings of Taunton brick and Rockport granite—the main structure four stories high and nearly 350 feet long—were rising rapidly. The mill was expected to run 30,000 spindles and 704 looms in the manufacture of printing cloths. It was predicted that both Potomska and the Wamsutta Mill which preceded it would double or quadruple their capacity within twenty years. Wamsutta stock was selling at $124 a share. The explosive rise of manufacturing presented such heady prospects that it was almost possible to forget whaling's gathering decline.

Nor was all the city's luster economic. Culturally, it was smug, too, and with reason. Three New Bedford artists were present at a meeting of the Century Club in New York—Albert Bierstadt, William Bradford, and R. Swain Gifford—and all were represented in what the *Evening Post* of that city described as "the finest collection of paintings which has been exhibited during the present season."

The play *Pomp* was presented in Liberty Hall and the editor of the *Standard* commented, "Among the large audience, we noticed that Joseph Howard, Jr., of the New York *Star*, Charles Taber, of the New York *Herald*, and Mr. Austin of the New York *Sun*, were enthusiastic in their applause. Coming from the village of New York, they seldom have an opportunity to see how the thing is done on metropolitan boards."

But other evenings at Liberty Hall, namely those featuring lectures sponsored by the prestigious Lyceum Society—in which the Howlands had been active participants for years—raised other questions.

Lecturer James Parton set forth the view that "until lately, the vulgar were the common people, mechanics and laborers . . . but now, the old nobility has declined in ability and quality." He added, and one suspects it made the backs of some of his audience stiffen, "In this country, not much honor is conferred by any external distinction of wealth or office. We are getting suspicious of munificent public gifts,* which I think are at least of doubtful benefit. We have no society here, although some make great efforts to imitate that of the rich and idle of the Old World."

Probably the reporter, sensitive to the views of the Establishment, reflected the reaction of many listeners when he wrote, "The lecture, we fear, was a little tedious and Mr. Parton is unfortunate in having a very poor voice for public speaking."

Another speaker in the series, Anna E. Dickinson, addressed herself to the "constant struggle now going on between employers and employed" and she acknowledged the "pitiable and tragical condition" of the latter as "the question of today."

Miss Dickinson said, "It is no longer a question of generosity, but

---

* New Bedford was looking forward to full payment of a bequest by Sylvia Ann Howland, forebear of George and Matthew, expected to amount to more than a quarter of a million dollars.

of justice. There is a discontent among workingmen, as revealed by the organization of labor reform movements. However, it is an insult to common sense to assert that capital combines against labor and as for the trade unions, they are not democratic, but autocratic; they are for the benefit of the poor workman and for the injury of the good; their aim is not directed against the all-powerful capitalist, but against the weak and defenseless who are outside of the organization.

"Trade unions insist that all be paid alike and work the same number of hours, the skillful, and the unskillful, willing and unwilling; they deprive of all chance of getting work those who do not belong to their organizations or conform to the rules they prescribe."

She admitted there was need of legislation for factory operatives and other overworked laborers, but maintained that the mechanics did not belong to this class. She directed attention to the results of an experiment in North Adams, Massachusetts, where both the employer of Chinese workmen and the workmen who set up for themselves "on the cooperative plan" made large profits. This, she suggested, showed there was plenty of room for every man, even in New England. Such cooperation, she declared, would give every man what he was worth and that was all he deserved. The *Standard* reporter did not say what he thought of Miss Dickinson.

It was not customary to publish letters to the editor on page one. The fact that this was done once in 1871 under the heading, "The Rich and the Poor," suggests that the editor of the *Standard* was aware of some of the things that were wrong in New Bedford—rich New Bedford, bustling New Bedford, culturally flowering New Bedford.

The letter stated, "A short time since, a lady, between nine and ten o'clock in the evening, was followed and insulted by a lecherous fellow. Her cries brought to her assistance a night watchman, who took chase and succeeded in overhauling him, but, upon raising his hat, it was discovered that he was one of our richest citizens and he was allowed to escape.

"Now is this right? Is it justice? Had he been a poor man and worn a tattered jacket, he would have been arrested and shown up to the public. In too many cases are the rich allowed to jump over or slip through, while the poor are required to suffer the penalty of the law. This is manifestly unjust. If there are any rich libertines prowling our streets, seeking to destroy the innocence of poor girls, the quicker they are shown up, the better for all concerned.

"A large proportion of the inmates of the houses of ill fame have been made such by the men who live in palatial residences and wear fine broadcloth and are the very first ones to pass regulations to send to the penitentiary the persons they themselves have ruined."

The letter was signed "Justice."

And some of those who viewed what was happening, if not with alarm, at least with candid disapproval, included the community's intellectuals. In his vine-covered "shanty" at Brooklawn, in the northern part of the city, Daniel Ricketson, poet and historian, scion of Quakers, yearned for the return of a New Bedford village, with its honesty and innocence, that was already obliterated by opulence. Ricketson was a figure of cultural, if not economic, consequence; he was acquainted with Emerson, Alcott, and Garrison; to his "shanty" came Thoreau, with whom he corresponded.

A contemporary article in *Harper's* said of Ricketson, who wrote New Bedford's first history:

He is as shy as a partridge and not only lives somewhat of a recluse from men, but actually hides himself under a broad-brimmed slouched hat and within the charitable folds of a huge old-fashioned camlet cloak, even when you are walking or talking with him.

His avoidance of society is instinctive as a musician avoids discords and he has a humorous twinkling sarcasm in his treatment of those who seem to him sophisticated or enslaved by society. A black hat or dress coat affect him like the most ludicrous jests and the habit of stuffing good honest English talk with French phrases excites his utmost contempt. He declares that he should as soon think of larding a beef tongue with the fat of frogs. Moreover, he is very fond of insisting that civilization has half-ruined us.

He said, "I believe that in society, people put on their best clothes to come together and see each other eat. I presume, from what I know of society, that they do so. I should be very much surprised if they did not."

Ricketson was largely unimpressed by what lots of money, rapidly obtained, had done to the appearance of New Bedford.

"It is to be regretted that so many of our more costly houses are built in a style of architecture which, however imposing at the time of their construction, will not bear a judicious criticism. The form of the Greek and Roman temple, however, beautiful and classic, was never intended for domestic residences.

". . . in a climate like our own, any style of building which does not admit of the sight of those objects that are essential to domestic comfort is at once to be set aside. The temple form, from the necessity of concealing as much as possible the chimneys as incongruous, is particularly objectionable. The chimneys, in fact, are considered by all good architects, when properly built and located, as among the most expressive and ornamental features of a domestic residence and they have been, by some old writer, very appropriately called the 'windpipes of hospitality.'

"There are but few styles of domestic architecture of European origin adapted to the climate of our country, as well as to the wants and genius of our people. Those edifices with their rich ornamental work built of enduring stone, when imitated in perishable wood, are always disgusting to the eye of a person of taste and suggestive of weakness and decay.

"The style called the Elizabethan, however beautiful in the old substantial stone or brick houses of England, surrounded by their extensive lawns, parks and pleasure grounds, is simply ridiculous when built of wood and, as usual, with some of the most important details omitted or supplanted by an altogether different style and placed in a lot of perhaps not more than half an acre of ground.

"Let us adopt a less ornate style of building than that which appears now to be so much in fashion. Every house, no matter how small and humble or how large and imposing in appearance, should have a *home* look; and if this idea is kept in mind instead of the present glaring, staring, illy-constructed edifices, we shall witness a harmony of effect which all must admire . . .

"The churches or meetinghouses, with one or two exceptions, are too outrageous to attempt a criticism upon . . ."

Of course, what Mr. Ricketson was *really* protesting was not this pillar or that gable specifically, for he acknowledged that "a very little alteration, and that generally in the roof, would render many of the at present distasteful, though costly mansions of our citizens beautiful residences."

What bothered him, and others of the thoughtful, were the things that New Bedford had become, that it wanted, that it was convinced were desirable and attractive, even necessary, now that it had become suddenly, unbelievably rich.

# III

# Into High Latitudes

# 10.

# The Whaleship and the Whale

Ships are but boards, sailors but men.
—*The Merchant of Venice*,
WILLIAM SHAKESPEARE

Never, in all of man's history, has there been anything comparable to whaling in terms of what it demanded of those afloat who pursued it, or of the vessels in which they sailed.

There is first of all the whaleship. It varied considerably in size and shape, depending on whether it was built for the job or converted from something else. The Howland brothers' *Rousseau* was originally a Philadelphia merchantman and old George had some of her upper works rebuilt to meet the demands of whaling; the *Martha II* was a slaver, and the bark *Perry*, presumably named for Commodore Matthew Calbraith Perry, was formerly of the United States Navy. Obviously, whaleships like *Concordia*, expressly designed and built for the industry, were more efficient. But even among such vessels, there was considerable variation.

Whaling vessels were roughly from three hundred to five hundred tons and some were fast sailers, especially in the later years of the industry, but many were not. An average ship was about 125 feet long; a good many suburban house lots are as long as that in one dimension or both.

Mostly, they were three-masted and square-rigged, that is, carrying yards on at least two of the masts. Being principally intended as floating factories and warehouses, the architectural emphasis was on strength of construction, cargo space, and ease of handling with only a small percentage of the crew aboard. It often happened that all of the whaleboats were lowered to chase whales with no one left aboard the ship but the cooper, as shipkeeper, and two or three others to help him as she lay hove to under shortened sail, or jogging "off and on," keeping the boats in sight and reasonably close by. Most ships carried a complement of about thirty. Their rigging was practical, strong, and not elaborate; its replaceability at sea was important, and this same consideration for self-sufficiency prevailed throughout the operation.

Officers and petty officers lived and slept in the after section, the remainder of the crew, forward, in the forecastle. Amidships was the manufactory, where whales were translated into oil and bone. The galley, or ship's kitchen, was aft, and every vessel, in addition to a cook, carried a cooper and carpenter.

When the ship sailed, with the prospect of a three-year voyage or longer—even though it was planned that she would be reprovisioned from time to time at way stations—what she carried reflected the intention that she must take care of herself, whatever happened, ordinary or otherwise. Belowdecks, she had quantities of flour, beef and pork; beans, rice, dried apples, salt codfish, coffee and tea, potatoes, onions, raisins, and vinegar, much of this stuff stored in tight casks because there was no way of knowing when everybody and everything might be exposed to a dose of salt water, from above or below.

The vessel was supplied with staves, called "shooks," and hoops for making casks; with sheets of copper sheathing for her bottom, to replace whatever was reachable that might be torn off by ice or weather; with spare spars, miles of cordage, and yards of canvas, to make her shipshape again after the gale's work was done; with medicine, paint, tobacco, twine, and nails, and a half-hundred other items that had to be thought of long before they were needed.

There was, moreover, a matter of official sanction involved. Implicit within this federal approval of the whaling voyage was acknowledgment that these ships sailed the world over and that, in so doing, they represented the United States of America. The document authorizing a voyage of the New Bedford bark *Java* measures about a foot and a half by two feet; it is signed by the secretary of state, bears the seal of the United States and in English, French, Dutch and Spanish, it states:

Franklin Pierce, president of the United States of America, to all who shall see these presents, Greeting:

Be it known, That leave and permission are hereby given to George W. Raynor, master or commander of the Barque called Java, of the burden of 291 and 70/95ths tons, or thereabouts, lying at present in the port of New Bedford, bound for Pacific Ocean, and laden with provisions, stores, and utensils for a whaling voyage to depart and proceed with the said barque on his said voyage, such barque having been visited, and the said George W. Raynor having made oath before the proper officer that the said barque belongs to one or more of the citizens of the United States of America, and to him or them only.

In witness whereof, I have subscribed my name to these presents, and affixed the seal of the United States thereto . . .

The bottom half of the document is directed to officials of foreign governments, opening with what must be one of the most all-encompassing salutations in the history of protocol:

Most serene, serene, most Puissant, Puissant, High, Illustrious, Noble, Honorable, Venerable, Wise, and Prudent Lords, Emperors, Kings, Republics, Princes, Dukes, Earls, Barons, Lords, Burgomasters, Schepens, Counsellors as also Judges, Officers, Justiciaries, and Regents of all the good cities and places, whether Ecclesiastical or Secular, who shall see these patents or hear them read: We, James Taylor, Notary Public, make known that the master of the barque Java, appearing before us, has declared, upon oath, that the vessel which he at present navigates is of the United States of America, and that no subjects of the present belligerent Powers have any part or portion therein, directly or indirectly, so may God Almighty help him. And, as we wish to see the said master prosper in his lawful affairs, our prayer is, to all the before-mentioned, and to each of them separately, where the said master shall arrive with his vessel and cargo, that they may please to receive the said master with goodness, and to treat him in a becoming manner, permitting him, on paying the usual tolls and expenses in passing and re-passing, to pass, navigate, and frequent the ports, passes, and territories, to the end to transact his business where and in what manner he shall judge proper. Whereof we shall be willingly indebted.

Thus did the business of getting ready for sea proceed. Sometimes the whalers left singly, sometimes in company, but the process was the same for each of them—a bonding of essential components, animate and inanimate, into an effective apparatus. Sound, tight hull; able crew, the more experienced the better; adequate provisions, and equipment good enough not to waste time or lose whales.

Thus was created the restless, ready ship, whose departure awaited only the moment of tide and circumstance.

Finally, the day came when they trundled the little foot-pedal organ in a wheelbarrow from the Seamen's Bethel on Johnny Cake Hill down the cobbled ways to the dock, and the chaplain held the service at the caplog for those leaving, and especially for those staying.

Then the ship was away, headsails and courses filling with the southwesterly, butting out of Buzzards Bay and bound for all over and in between, bubbles under her determined forefoot and moving faster than ever in her life, at least to those aboard and those who stayed behind, both suddenly conscious of the long time ahead and the great distance between.

When the land is hull down, gone into the wake and beyond seeing any more, the whaleship is its own world. It not only endeavors to feed, doctor, repair, protect, and maneuver itself; it also possesses its own god and government. Landsmen can have no concept of the responsibility this placed upon the ship's master and, to some degree, upon his officers for years on end. Obviously, all masters were not equally qualified to assume this responsibility; some never even tried and would have failed largely if they had. So there were, on occasion, bloody mutinies, unspeakable persecutions, and other dark matters of many kinds, now covered by time, and better so.

Yet the wonder of it is that the system worked so well so much of the time and it is a credit to the master mariner that it did, that he was, in fact, generally more successful than otherwise. In part, this was because he was an unusual breed—any number of his contemporaries would have refused his job, even if they had been qualified; in part, it was because he rose necessarily to the demands of a task that asked him to be virtually everything, twenty-four hours a day, and with no overtime.

If a man's foot is crushed between a whaleboat and the side of the ship, somebody has to decide, quickly, whether any of it can be saved without risking the man's death of gangrene, or whether it has to be sawed off, while somebody pours rum into him, and somebody else tries to hold him still on the after-cabin table. Neither squeamishness nor useless philosophizing about who reasonably has a right to make his brother a cripple or a corpse must be a deterrent. The worst thing of all, as in most aspects of whaling, is doing nothing.

If the bottom is dropping out of the glass—that is, if the barometer is falling—somebody has to decide, before it is far too late to have a choice or make a decision, whether to shorten sail, retain the present course, heave to under bare poles, run for a lee, if there is any within reach—whether, in other words, there is potential danger to the ship and all aboard, and if so, what can best be done about it that might (not "will," just "might," for there is no such certainty as "will" at sea) prove practicable.

If, among the thousands of miles of uncharted waters through which the whaleman sails, guided by who knows what instincts, half-remembered experiences, preachings of some mentor, smell, split-second observations, and sometimes fear and desperate hope, something smashes the essential hull that keeps sea from devouring man, somebody has to figure out how, if possible, to stop the leak before the ship sinks. And if keeping her afloat is not possible and there is no one nearby to help—and often, there is not—somebody has to decide when it is time to abandon ship and how to go about it, so as to give each man aboard the best chance to survive. Add to this that some of the crew are bright and resourceful. Add, too, that some of them are sailors born and bred and some of them, especially in the early part of the voyage, are farmers, notion-counter clerks, and city fellows seeking adventure. The ship's master must balance the equation between the task's demand and the human resource as best he can.

In addition to all this, the master's owners have invested thousands of dollars in the ship, advances to officers and crew, equipment and provisions, and future drafts upon their accounts that he is authorized to make as necessary. The land-based stockholders in this whaling venture expect the captain to make, every hour and every day, decisions best calculated to preserve their expensive investment, to conserve the money and material involved to his best ability, and to provide them with a reasonable return. Some even expected an unreasonable return.

I have known, directly or otherwise, several whaling masters who were, I am sure, reasonably comparable to their immediate predecessors, the captains of the Arctic fleet of 1871. To suggest that these men were not distinctly individualistic would be ridiculous; they varied from virtual illiterates who had worked their way aft from the forecastle through courage, hard work, good seamanship and fiercely effective pursuit of the whale (and who had to ship every voyage a competent mate to navigate because they could not) to genteel, soft-voiced mathematicians, one of whom said to me with quiet confidence, "Give me a dollar watch and I will take any ship anywhere in the world."

Yet there was this in common about them—their humor, subtle or brash, concerned itself most often with man's smallness and temporariness in the universe. They understood, without actually saying so, that death waited in the wings every day, yet understanding it, did not concern themselves with it. "Not one damn," as one of them said to me. Understanding what man could not do, they nevertheless had great faith in what he could; they possessed extraordinary self-confidence, unshatterable nerves (I do not remember one with facial or other mannerisms; they sat and stood with the calm of eternity), and they could, at a moment's notice, tell you at least one reasonable way of doing almost everything on earth with which they had ever had contact.

In almost every way, the whaling masters repudiated the landsman's concept (especially that of the journalist and novelist of their times) of what they were. They were scornful of what these people had to say about them and their scorn included Herman Melville, whom they knew less as a writer than as a ship-jumper. They did not understand at all what he had written in *Moby Dick*, or why; they had a vague notion that he was a homosexual, and they believed he had purposely drawn an unbecoming, perhaps even indecent caricature of what they were and did.

These men seldom used profanity and I have never heard one, even those from the most modest backgrounds, revert to obscenity. Their morality was of a practical kind, but inflexible; they took it for granted that they had an obligation to decide what was right or wrong and that nobody (God included, whom most of them respected) would overrule them. They were, of course, not always good-natured, yet they were predictably consistent in their responses. This is the kind of temperament with which you can live aboard ship; in other words, if the cook knows that the Old Man hates his beans overdone and will fire beans, plate, fork and all at the galley if they are, well then, the cook is a fool if he overdoes the beans twice.

Considering all this, even if he was well-paid (and he often wasn't, through no fault of his own), the whaling captain was not overpaid.

To a degree, his officers shared many of these burdens and usually had, even more than the master, the responsibility of taking the whale once it was "raised" or sighted. The profit-sharing aspect of whaling, as well as the Old Man's disposition, drove them to do the best they could. They also often had to make decisions of grave importance with no time in which to do it, and their working-day lives straddled a balance between common sense and derring-do. They were more aware than anybody else that if they took a risk and it resulted in a dead whale, they were the richer in pocket for it and might get a good cigar from the captain as well. ("Compliments, Mr. Smith. Theftily done.") And if the risk did not result in a dead whale, but in damage to life, limb or property, they would catch hell.

As for the foremast hand, the seaman in the forecastle, he was by nature or necessity a model of quiet endurance, there being no alternative. Consider excerpts from the journal of John G. Abbott, a young man of Montpelier, Vermont, who sailed from New Bedford aboard the *Huntress* in the spring of 1836. (This ship was one of the early casualties of bowheading, being lost in 1852 on Kaiaghiasky Island, Kamchatka). The portions of Abbott's journal are especially selected to reveal how much of what happened to a whaleman was no more than incidentally related to whaling.

On June 9 of 1836, Abbott and others went ashore at the Portuguese offshore island of Brava, in a whaleboat, to pick up fresh meat, fruit and vegetables. At four in the afternoon, they left the island, the boat being heavily laden with supplies, and started for the ship, which had been standing off and on, waiting for them. They rowed to windward for twelve miles and then, in Abbott's words, "We begun to halloo after dark, but could hear nothing nor see. The wind blowed hard and the sea began to run high. Some of the Sailors began throwing the Pumpkins overboard, for the water broke in too the boat. They was a-going to throw some of the Hogs but was kept from it by some of the rest. Some was for going to the Shore and some for the Ship but we did not know where either was. At last we saw a light. It proved to be in the Cabbin Windows of the Ship. We puled for it with all our might for a half an hour, hallooing all the time and they heard us and hove the main yards aback [to halt the ship's weigh, so the boat could overtake] and we came up and got on board some tired and wet."

On August 3, a "small black Cloud" was observed rising. It soon spread "all over the Horizon. There was hardly a Breath of Wind and we began to take in all Sail. Soon we heard it roar and the Captain ordered all Sail to be Clued up.

"The Lightning Flashed all over the Hemisphere. It streaked Down our main-topmast fore and aft stay and went off without doing any Damage. It left something that appeared like a ball of Fire on our main truck at the top of the Mast which stayed about three minutes and went off presently.

"The wind struck the Ship. I was at the Hellum. The Captain took hold with me and we put the Hellum up and kept her off before the Wind under bare poles. She went off twelve miles an hour and ran about four hours and it lulled a little. My hands were not a little bruised as they were exposed to the Hale Stones, which fell as large as Walnuts."

Abbott went ashore at St. Paul's Island on August 17. He found, "There is a hot Spring there about a rod from the shore that will Boil Fish. We caut some and put intoo it and they were Boiled in a few minutes ready for eating. We ate some of them; they were very Good."

Having come, late in September, to the "Land of New Holland," he found that the natives threw a spear very accurately at the "Distance of 12 Rods, ate Cangeroo, Fish, Snakes and Worms, worshipped the Moon, and were horribly afraid of a Gun." His relations with one of the few white settlers there, however, were neither as objective nor as pleasant.

"The Captain asked mee if I would go with him [to the settler's] and play a few tunes on the Violin. When we got there, Mr. T—— took us into the best room he had got. I played two or three tunes. Mr. T—— brought on some wine and invited the Captain to drink and did not aske me although I had my mouth made up to say no, but he did not give me the chance. He supposed I was the Captain's slave seeing that I was pretty black. [Presumably from sun and weather.]

"However, I laid down the violin and took my hat to walk out and take the air and they got no more Fiddleing. Afterwards, the Captain said he thought it strange that he did not ask me to Drink."

At New Holland, they raised the ship's bow by filling casks on deck aft with water, in order to get at a leak forward and repair it. They also went ashore and shot parrots, got a mahogany anchor stock and obtained a monkey. "They [the natives] know little about Choping here," Abbott recounted, "and we bet Three Dollars against the monkey that one of our men could chop a large mahogany tree Down in one hour. A man by the name of Hull went to work and choped it Down in seventeen minutes and so we took the monkey."

Twice in twenty-four hours on February 26, off the island of Lord Auckland, all hands were turned out—once at night—to get into the boats and tow the ship offshore. Light air and strong currents had set her to within "a ship's length of the island."

Here, they obtained several ducks for food, "knocking them over with clubs, for they had not seen a Man before." They also shot seals to replenish the larder but, in a sense, replaced what they took by carrying "a Shore two gees and four Rabbits to colonize the island."

They stayed some time in this locality, being kept awake on occasion by the "barking of seals," and experiencing several "snoe storms." At length, they prepared to get underway and all hands turned to to heave up the anchors (three of them). By midnight, they had two of them up, after working continuously from early morning.

"The Captain," said Abbott, "was a-going to make us heave up the other ancor that night and he sent some butter to us for supper, to encourage us to do it after working hard all day, but the wind died away, so that we could not sail out if we had hove it up, so we got the butter for nothing."

When they did get the anchor up, the fall (hoisting tackle) parted and in getting the anchor onto the cathead to stow it, they dropped it irretrievably overboard. "The Sailors thought it would lay hard on Sarah's Stomack," said Abbott, "that being the name of the bay." Although the *Huntress* went to sea "sloe and steddy," all hands were "dizzy," having been in the bay at anchor for three months. (They had pursued the whales close inshore during this period, using only the boats.)

A day out of St. Catherine's, they sighted a boat of about eight tons, dead to leeward. The captain remained aloft with his "spye glasse" as they ran down to speak the boat, imagining it to be from some "recked ship.

"We got within a half a mile of her and the Captain sung out, 'Hard down the hellum and brace the yards up sharp,' and he came Down as fast as he could. The boat was laying by for us all the time until we hauled on the wind and then they made sail after us.

"The Captain said he see thirty or forty men in the Boat and that it was a pirate. We had a good breeze and kep on and they after us. They follow us for about five hours but we gained on them all the time and they thought it was best to give up the chase. We hoisted our flag but they shoed none and finally they tacked and stood in for the land." *

So much for the background. Now on with the business of catching whales.

Next after the men, there is the business of the gear and this brings up the whaleboat, which is carried aboard the whaleship. Most ships had four or five and a couple of spares. It is in the whaleboat that a crew of six goes to capture the whale, if it can be done. A whaleship is the only vessel that can lower all of her boats simultaneously and this is because each boat's crew is a self-sufficient entity in this and many other facets of the operation. It is desirable to hurry when lowering; the whale isn't likely to wait, and what is more, he often gallies at certain manmade sights and sounds. If he does, he may disappear beneath the sea or take off for far parts at a rate faster than you can keep up with.

The whaleboat is a poem of its kind, precisely designed on the basis of generations of experience to do what it is expected to. Approximately thirty feet long and six feet wide amidships, it is pointed at both ends; its carrying capacity compared to weight is astonishing; its simplicity of construction makes it easy to repair; its rise of bow and stern makes it ducklike and dry in a heavy sea, and its sharp, clean lines make it fast and maneuverable. In a chase with a whale, it is possible to make five miles an hour in a whaleboat under oars alone —as long as the crew can keep it up, that is.

Five of the crew pull oars; the sixth steers. The oars are thoughtfully counterbalanced; the harpoon (most forward) and after oar are fourteen feet long; the tub and bow oar, next inboard, are sixteen feet, and the midship oar, in the center of the boat, is eighteen feet long. These are so placed that the two shortest and the one longest

---

* Even though Abbott was concerned with the early nineteenth-century, incidents of whalemen's difficulties with natives persisted for many years. In 1859, the ship *Twilight*, Captain Sylvester Hathaway, was lost at the island of Hiva Oa, in the southeastern Marquesas, and he and his crew "had trouble with natives, but were protected by a missionary residing there." In 1862, the whaleship *Reindeer*, commanded by Captain Raynor (whose ship's papers, when he was master of the *Java*, were quoted earlier in this chapter) was "attacked by Arctic natives."

pull against the two sixteen-foot oars, which preserves the balance when the boat is worked with four oars, the harpooner at that time stowing his oar "apeak" and being busy with the harpoon. The boat is steered by an oar twenty-two feet long, shoved through a grommet in the sternpost.

All of the pulling oars lie between tholepins of wood in the gunwale. All sound of the oars is muffled by well-thrummed mats carefully greased.

In the whaleboat, in addition to the crew of six, are two "live" harpoons, ready for use; two or three spare harpoons, two or three lances (barbless blades for killing the harpooned whale), all with their sharp heads in wooden sheaths so that nobody will fall into one and become a casualty. There are also two line tubs, in which are coiled three hundred fathoms (1,800 feet) of hemp line, arranged with every possible precaution against kinking in the outrun. Many a whaleman has died because of a fouled line; if as it snakes from the tub, hauled by a hurt whale off for the horizon, it kinks and catches an arm or leg, the man goes overboard in an instant. Often he is never seen alive again.

To this gear, add a mast, spritsail, hatchet, and sharp knife for cutting the line to the whale if he threatens to haul the boat down with him or smash it to splinters; water keg, candles, and lantern (in case the boat can't get back to the ship in daylight), compass, bandages; waifs on poles, for flagging the carcass, so it can be spotted at a distance; a fluke spade, for cutting a toggle slot in the whale's tail, so that he can be towed back to the ship; a drug or dragging float (to slow him down when he runs); and a boathook.

There is obviously no place for a clumsy man in all of this.

To understand what the boat and its crew are pitted against, bear in mind the dimensions of the average bowhead whale, whose head is a third of its bulk. Consider his mouth as having a three-hundred-barrel capacity, all but fifty barrels of this—the amount of a mouthful of food-filled water—being taken up by his huge tongue. This mouth, although narrowed sharply at the top, is nearly ten feet wide at the bottom; twice that in length, and more than fifteen feet high. Plenty of room for Jonah.

The fat white lips are four feet thick, and so is the throat. The lips and throat of a two-hundred-and-fifty-barrel whale should yield sixty barrels of oil and with the supporting jawbones, weigh twenty-five thousand pounds. A bowhead's tongue, weighing five tons, has yielded as much as twenty-five barrels of oil. When the creature feeds, his lips are spread to as much as thirty feet and he strains the minuscule organisms that are his meal from a quarter-mile of ocean before he forces the water out of his mouth and downs the brit on his bone-hair strainer. His tail is often twenty-five feet broad.

Barrels of blood more than one hundred degrees warm pour through the massive pipes of his circulatory system, the largest a foot in diameter (as is his respiratory canal), driven by a heart itself as large as three barrels.

The whale's bone, which usually runs from eight to ten pounds for each barrel of oil yielded (although the ship *Sarah Sheafe* took a bowhead in 1857 that produced one hundred barrels of oil and three thousand pounds of bone), was at this time used in the manufacture of whips, parasols, umbrellas, dorsers, caps, hats, suspenders, neck stocks, canes, rosettes, cushions for billiard tables, fishing rods, divining rods, bows, busks, probangs, tongue scrapers, pen holders, paper folders and cutters, graining combs for painters, boot shanks, shoehorns, brushes and mattresses.

And this is how whaleboat and whale come to an encounter, once the lookout (preferably an Eskimo, who can spot a bowhead when nobody else can), leg hooked over the fore topgallant crosstrees of a whaleship, sees the spout or chunk of black back that indicates prey in the offing.

If possible, the ship works to windward of the whale before lowering; it is far easier and faster to approach him down the wind than up. When the master feels his position is at its best, and bearing in mind the need for haste, the vessel is luffed up—that is, pointed so sharply into the wind's eye as to lose way. Slowed, the boats may be lowered from the vessel more easily and safely.

Four members of a boat's crew man the davit falls by which the boats are lowered into the water; two members, one bow and one stern, are lowered with the boat; they unhook the boat from the falls after the other four have slid down to join them.

Now the boats are away; they set their spritsails, and the Old Man is aloft, and squirming impatiently. The whale has sounded, gone down; he'll be up in twenty minutes and the boats, eager to strike this rising, do not crowd but spread out in order to improve their chances.

The whale's upcoming is announced by a quarter acre of turbulence unlike anything other than the sea breaking from the bottom. The phenomenon arises from the same general causes in each case— a forceful upwelling, irresistible and of massive proportions. And out of the broadening ripple and roil, the gathering swirl that makes the mouth dry, the heart pound, knowing that the monster is close and closing, there—just there—he breaks the surface in a roaring and pouring of seawater, shiny black and wet, and with the damp, hoarse emptying of the great lungs.

Upon command from the boatheader in the stern, the harpooner is up; he stands in the bow, his leg braced in the timbered half-circle called the clumsy cleat, to steady himself. He holds poised a six-foot sapling pole jammed into the socket of a thirty-inch iron shank on the end of which is a toggle iron, pointed to find its sharp way home, hinged to tip like the crossing of a "T" so it won't pull out.

Now the black back is close aboard.

Contrary to the layman's notion, harpooning is less a matter of hurling than of brute strength and nervelessness—in order to be most effective, you have to be close enough to the whale to be in danger. "Give it to him," says the boatheader quietly, and the harpooner bends to his work. What happens here now in something like fifteen seconds is part clockwork and part ballet without music; success, and perhaps even life, hang upon the precision and timing.

The harpooner bends, arm, shoulder and back in graceful coordination, and drives in the first iron. "Clear to the hitches," says the boatheader with satisfaction. In the same instant, even as the harpooner reaches for the second iron (to sink it in the whale if he can, or heave it overboard for later retrieving if he can't), the spritsail is doused, thirty feet of "box line" is thrown overboard as soon as the whale is hit—he must take up this slack before line starts running out of the tub and it will give you time to get ready for whatever happens next—and aft, a turn of the tub line is taken around the loggerhead, a vertical post, to make it harder for him to run out the line.

Run he does. With the first terrible spasms of fear and agony through his great body, he swims for the edge of the icepack, knowing that if he can get under it, he is safe. The line that he has ripped off in his initial lunge must be retrieved. In bowheading, because the open water is limited, it is essential to haul up on a whale as fast as possible, to hit him again with a killing blow before you lose him in the ice where you cannot pursue him.

Frustrated in his running, hard-pressed in the effort to escape his tormentors, he smashes his broad flukes upon the water, sending up a shower of salt spray, and he sounds, bound down into the cold green waters that have, always before, offered him sanctuary. The water is shallow; soon the outpaying line halts, and now is the waiting time for man. (For the whale, there is almost no time left and down there below, he is hurting.) Soon, in the boat above, they feel him coming up on the line; they retrieve it, coiling it, not in the tub, but in the boat this time and as carefully as before, for it may go smoking out once more if the whale feels up to another run for freedom.

But the irons were deep in this one. When he comes to the surface, the spirit is draining. The boatheader changes places with the boatsteerer; it is the former's prerogative and duty to kill, to find the life deep within the monster with the lance—either hand-thrust or driven with a black-powder charge in a shoulder gun—and to extinguish it.

They paddle close, and the whale lies there, without will or vitality to avert its death. There is, abruptly, as they draw alongside, one last moment of magnificent protest, one last sweep of the wide-ranging flukes, and a shower of water drenches men and boat. But in the same instant, the lance finds its way over the shoulder blade and downward, downward, and the great shower of blood comes. He rolls fin out, with one final, awful pumping—all that sustained him through leagues of silent voyaging in the depths is now denied him for all time.

Still, every man in the boat, soaked with salt water and blood, knows that this is not simply a dead whale. It is $10,000, granted any kind of reasonable market, and everybody aboard gets some of the dollars. What they are thinking about right now is whether the lance punctured his lungs; apparently it did not, because there is no rush of air bubbles in the water around him. Down over the shoulder blade is the best way to kill a bowhead in a hurry, but if the thrust deprives him of the buoyancy lent by his lungs, he will very likely sink, carried down by the ton of whalebone in his mouth and you have had all your effort for nothing.

He floats, and that is good. The Old Man is down from aloft now and standing on the sheer pole, waving his hat. That means, "Well done, lads; hustle him alongside, cut him in, and get after another. Time is money."

Because the ship is close by (and it usually is, in Arctic whaling) a crewman takes the line from the dead whale to the vessel and puts a turn around a loggerhead on deck. All hands aboard tail on the line. The trick is to let the roll of the vessel pull the carcass alongside. When she rolls, one gains some, then takes up the slack quickly, and waits for the next time. The whale is drawn in on the starboard side, the head aft and the fin in front of the gangway. With a twenty-foot boathook, an egg-shaped float to which a line is attached is shoved under his flukes. When it pops to the surface on the off side of him, the crew hooks it back to the ship, fastens a chain to the line and, with the line, hauls the chain around his flukes and back into the vessel's chain pipe. One end of him is now fast to the ship.

Then another chain is made fast around his fin (it's bigger at the outer end than it is at the base and the chain cannot slip off). This is how the whale is managed during the cutting in.

The cutting stage hangs out over the water in front of the ship's gangway; it is a trapezoid of four planks, the broader base against the vessel's side. The stage has around its three outboard sides a row of iron stanchions with a lifeline rove through them to keep the cutters

Bark *Oriole* cutting in off Point Barrow, c. 1870. *Courtesy of The Whaling Museum, New Bedford, Mass.*

from falling overboard while they work, for the planks get slippery with blood, grease, and seawater.

With long-handled cutting spades, honed to razor sharpness on the ship's grindstone, the blubber is cut through, down to the "lean," circling the whale from the mouth toward the tail, in corkscrew fashion. When the fin chain is hauled on, the carcass is rolled over, and the "blanket piece" of blubber that has been cut along its edges, tears off, assisted by some more prodding of the cutting spade along its under side.

Rolled quarter over, the whale's lip comes up and the cutting fall (a contrivance of oversized blocks and rope hung from the mast and designed to raise and lower heavy weights) is hooked into it. Forward, the windlass is manned, to do the hoisting. The lip is cut off, hauled aboard, and dropped into the blubber room (the space between decks from the mainmast to the forecastle).

Now the blanket piece is started again and the whale is rolled halfway over. The throat is up; it is cut off, and dropped into the blubber room; the other lip is rolled up to the surface and removed. By this time, the blanket piece, which has been unwinding all this time (like a long peeling coming off an orange), is getting long enough to be unwieldy and the pulleys of the tackle (preferably pronounced "tay-ckle" by whalemen) hauling it aloft are close to "two-blocking," that is, coming together so that the strip may be hauled no higher.

At this point, the blanket piece, roughly fifteen feet long and six feet wide, is severed from the whale at its lower end and dropped into the blubber room; the cutting fall is lowered to the carcass, and another hold secured with the blubber hook, preparatory to pulling off another blanket piece.

This is a critical time because the crew is now ready to cut off the upper jaw, in which is located all the whalebone. The whale may have as much as three thousand pounds of bone, very likely worth at least two dollars a pound, and the bone is worth so much more than the oil in him that bowheaders many times did not bother to try out the blubber at all. One clumsy move and the head, once severed, may go to the bottom.

A hole is cut between the scalp bone and the tough blubber around the spout hole. Through the hole is shoved an eight-foot iron needle, which is double-eyed; it can be pulled through with a line on the front end and carries a line in the after end as well. With the line rove through the hole, a chain is attached to the line, pulled through the hole after it, and the chain is hooked to the cutting falls. With an axe, the whale's backbone is chopped nearly through, close to where the blanket piece was started; then, with a jerk of the hoisting tackle, the weight of the carcass breaks what is left of the backbone and the whole head is hauled on deck.

Meanwhile, the crew continues to peel the blanket piece off in strips until they get close to the flukes. There, the backbone is disjointed and the final haul on the tackle brings on deck the whale's tail, with the last of the blanket piece. The orange is peeled. The carcass, now bearing only lean meat and no fatty overcoat, is set adrift. It may either sink or float, depending. Sometimes, for a change of diet, the whalemen cut off some tenderloins. Whale meat is as good as stew beef and sometimes better.

The flesh is not "fishy," of course, because the bowhead is not a fish and the meat can be used very effectively in soups and stews, boiled, or put through a grinder and mixed with whatever suits your fancy, including hardtack, onions, and potatoes.

When the last piece of blubber comes on deck, it is customary for all hands to yell, "Hurray for five and forty more!" or "Old Hallett!" Captain George Fred Tilton, who said he had hollered "Old Hallett" more times for more years than he could remember, confessed that he had not the faintest notion who Hallett was or why the whalemen yelled it.

It is time to start up the tryworks, building a cordwood fire under a couple of huge iron pots, round on two sides and flat on two sides for easier stowage. The trypots, four feet in diameter and two and a half feet deep, are of about two hundred gallon capacity; they are enclosed in a brick ovenlike affair closest to the foremast under a roof called a hurricane house, a structure open on the sides from the bulwarks up.

Within a watertight box of plank built on deck, called a "duck pen," a floor of bricks is laid in sand. On this floor, set in mortar, is the supporting brick for the trypots and the walls that enclose them. When the ship is trying out and has a fire going to keep each pot hot, the duck pen is filled with water which circulates through the sand and around the bricks, thus protecting the wooden deck. Smoke is carried off through two copper chimneys, four to six feet high—one for each pot—that go up through the hurricane house.

In the blubber room in the 'tween decks (that is, below the main deck and above the hold, where the casks of oil are stored) the blubber is cut into horse pieces, something like three feet long, six inches wide and, in the case of the bowhead, sliced horizontally, to reduce its thickness by half, for easier handling.

These pieces go to the mincer, who cuts them into "bible leaves," so they will try out faster. The mincer's horse is of wood, rounded on top like a log; it has four stakes projecting upward from it to make the horse piece run straight. The mincer stands astraddle of the horse, wielding a tool like a broad-bladed drawing knife, with a handle at each end, bringing it down on the horse piece with a slanting, slicing movement. He does not cut quite through the blubber, stopping his stroke at the gristle called "white horse," so that the piece, scored deeply, in the old-fashioned manner of pork for beans, will hold together.

The chunks of blubber then go up on deck and into the trypots, and the smoke rises black and roily through the chimney pipes. Everybody, near and far, can see that you are "boiling," and very soon, there is a greasy film over almost everything and everybody engaged in the operation. After the oil is boiled out, the hard and brittle pieces to which the blubber has been reduced, pressed dry of the last drop of oil, are thrown into the fire for fuel and cordwood is not burned again until next time. The blubber burns with a vengeance.

If the night shuts down while the ship is trying out, blubber scraps are placed in a metal basket called a cresset—even as it is on the Colossus of Rhodes. The contents are set afire and hoisted overhead, where they give ample light to work by. Some say they burn better and brighter than pine knots.

With a long-handled bailer, the oil is taken out of the trypots and poured into a cooler on deck, the latter being a rectangular tank of sheet iron that will hold two to three barrels. A leather hose, with a valve controlling the flow, leads from the cooler into the vessel's hold and with it, casks below are filled through their bungs, a cask holding between thirty and thirty-three gallons.

The slabs of overlapping whalebone are chopped out of the whale's head with a cutting spade and, in between whales, when the crew is not occupied, the slabs are scraped of gum—"knocking off the oysters" in the whaleman's phrase—and thus cleaned, they are bundled and stowed away. If there are Eskimos aboard, they relish the gum shavings from the bone which, according to those who have tried them, taste like decayed raw peanuts.

Whaling is, in short, like nothing else. To deny it its poetry, drama, and pathos is to ignore the struggle—man against man, man against creature, man against nature—of which its fabric is woven. To deny its uncomfortable, wearisome, boring, distasteful hard work is to turn one's back on reality.

Yet somewhere in between, there were such matters as sunsets and satisfaction for those responsive to either. It is no wonder that, like the parable of the blind men and the elephant, the whalemen's versions of what whaling was like varied as night and day; it is no wonder that some of them loved it and some could not stand it.

# 11.

# The Sandwich Islands

It was evident at first view of the houses [of Honolulu in the late nineteenth century] that the old was still struggling with the new, for regular New England cottages with green blinds and stone structures were mingled with mere huts of straw . . . My first impression on entering the place was that all the true natives had gone off in their canoes, for we at first encountered none but Chinese and other apparently Asiatic people, and then Portuguese, Frenchmen and a few English and Americans, but . . . we saw that those were only the advance guard of porters, small traders, agents for this and that, idle sailors and the general flotsam and jetsam of a port in the tropics which is the common resting place of rovers of all nations.
—JOHN F. WILLOUGHBY,
*Hawaii, the Pearl of the Pacific*

Like New Bedford, Honolulu—the Arctic whaleman's last stop for outfitting, provisions and recreation before sailing for a season in the north—exploded into sudden prosperity on an economic diet of blubber and bone and related commodities; like New Bedford, Honolulu suffered as a result.

The touch of whaling upon a community was increasingly a harsh touch, especially as the financial uncertainties of its later years tended more and more to attract gamblers at the investment level and rascals in the forecastle.

When Hawaii's sandalwood gave out, leaving only a small trade in sugar, the whaleman filled the breach; as early as 1823, there were forty to sixty whalers calling at Hawaiian ports for refitting and provisions every year. Sailors spent their money on rum, women, and native wares to take home for souvenirs. Some New England merchants eyed the commercial bustle there with envy and set up business branches in the islands. More whales, more ships, more oil, more money—around and around the busy wheel went. From 1836 to 1841, 358 whalers touched Hawaii and imports from the United States amounted to a million dollars. Honolulu, a cluster of grass huts, three or four stores, and a native population of about twenty-five hundred at the beginning of the nineteenth century, soon doubled in size and counted among the newcomers more than six hundred foreigners, called *haoles* by the natives, come to make a fast dollar.

Between 1840 and 1860, an average of four hundred ships sailed into Hawaiian harbors yearly. Local residents boasted, "You can walk from one end of Honolulu harbor to the other, ship to ship, without getting your feet wet." From 1851 to 1860, 4,420 ship visits were recorded in the marine register at the port and 14,138,714 pounds of whalebone and 17,661,446 gallons of whale oil were reshipped on merchantmen bound for the States.

The regular visits of the whaling fleet had a major impact in shaping the history of the islands and their people; the result was great financial gain and equally great moral loss. In the early years during which Hawaii became an outpost of New England, sometimes as many as six hundred whalemen would be ashore at a time and it was a running battle between the missionaries and the *haole* merchants whose profit derived from encouraging sailors and natives alike to indulge in whatever.

Above the wooden belfry of the Seamen's Bethel at Honolulu flew the flag of salvation for the lonely sailor far from home, to whom the Reverend John Diell and his wife distributed Bibles and spelling tracts, the latter because many men of the Yankee fleet were illiterate. Below the belfry, in the crooked streets of the *haole* district (four hotels and nine grogshops), the word was, "There is no God this side of Cape Horn." The community treasury was swollen by fines for disorderly conduct; the riot of '52 was so serious (drunken seamen set fire to the police station and tried to burn the whaling fleet) that resident foreigners formed the Hawaiian Guards to quell violence in the streets; in the peak year of 1856, when 596 ships came to port, the Guards worked night and day to protect Honolulu from wanton destruction.

On the other hand, the arrangement possessed assets and attractions for all the principals. The shipmaster might rather have gone into Lahaina where there were no saloons and thus no problem of social disorder; moreover, its port charges were lower. For that matter, San Francisco had better facilities than the ports of Hawaii, especially, transport links. But Honolulu had no "land sharks" seeking to shortchange both sailor and captain; by contrast, because of his economic importance, the master of a whaleship was the biggest frog in the puddle in Hawaii, and the agent took such good care of him ashore, and the Hawaiians took such good care to please the agent, that the captain's stay in the islands was more or less a holiday.

From the islander's view, it was big business. By 1843, the average annual value of American property touching at Honolulu, including the outfitting of whalers, amounted to $1.2 million. For all the islands that year, including cargoes of oil, it involved a $4 million gross and more than two thousand sailors with money in their pockets.

Merchants, warehouse owners, shippers, and shipyard workers all benefited from the Americans' visit twice a year, and so did the islands' agricultural community. Cattle were driven to the ports for slaughtering; there was a big demand in the fleet for firewood; the farmers planted the temperate mountain slopes heavily and sold to the ships tons of white and sweet potatoes, pumpkins, bananas, molasses, onions, sugar, coffee, coconuts, breadfruit, taro, cabbages, oranges, pineapples, melons, as well as turkeys, hogs, and goats.

By 1871, both the vice and the volume had dwindled. Annual arrivals of whaleships had dropped to about fifty; in a twenty-year period oil brought into the island ports for transshipment had decreased from 375,000 barrels to 20,000. There were fewer white sailors ashore, far more law and order, the natives were better adjusted to *haole* ways, and the *haole* residents themselves had come to contribute stability to the community. But even though whaling was in decline, oil transshipment was still important to Hawaii (1.8

million gallons as late as 1869) because longer voyages in an effort to make a paying trip meant more transshipment, and because the Panama Railroad, opened at midcentury, operated a cargo ship service at its eastern terminal and emphasized its facilities for east-bound oil shipments to New York.

Certainly one factor that continued to attract the Yankee whaler to the islands was the availability of manpower. The Hawaiian was a sailor by instinct, a superb boatman, and the average whaling master was eager to sign him on not only because of this, but because he was also likable and industrious. The typical native was friendly, even-tempered, and generous; he came from a society that knew no class or race barriers and he was, more often than not, a respecter of character, intelligence, and, because of missionary influence, middle-class morality. (As Mark Twain wrote of the Hawaiians, "They all belong to the church and are fonder of theology than they are of pie; they will sweat out a sermon as long as the Declaration of Independence. The duller it is, the more it infatuates them . . . Sunday Schools are a favorite dissipation with them. They never get enough.")

Of the approximately fifteen hundred who sailed north for the bowhead season of 1871, about half were Kanakas, a name that the New Bedford whaleman applied indiscriminately, and usually erroneously, to all South Sea Islanders. If a shipmaster could not spell a native's last name, he often recorded him in the ship's articles as John, Sam, Joe, or Henry Kanaka, or gave to him as a last name the name of the place in the islands from which he came.

It was common for ship's articles to be set up in double entry form to accommodate the two languages. On one side of the page, it read, "*O maku ka poe i kakau i ko maku maul inoa malalo ke ae aku nei makou a ke hoopaa nei makou ia makou iho e,*" and on the other side, directly opposite, "We whose names are hereunto affixed do hereby agree and bind ourselves to serve in the capacities set opposite our names."

How the islands and their people appeared to the voyaging American of the late nineteenth century who had sufficient learning and

interest to record his reactions—and many whaling masters did—is
evident from the following, written in 1898 by John F. Willoughby,
for the American Press Association:

From the center of the town [Honolulu] on, we struck it rich in the
Kanaka line. We saw them of every age and many shades of color and
in every stage of dress and civilization, from the new arrival from the
mountain highlands, clad only so far as absolute decency required, to
the cultured gentleman in blue coat, white vest, and truly immaculate
linen.

At every convenient recess in the side of the street was a group of
native women. With scarcely an exception, each wore a sort of Mother
Hubbard of blue cloth and whether walking, sitting or riding—and
many of them were riding, and always astride—they managed this
single garment with modest grace.

The streets were whiter than was agreeable and rather dusty. The
weather changed quickly. It began with dead calm air, replaced suddenly
by a strong, cool breeze. In three minutes at the furthest, the sky was
overcast and almost black; in ten minutes more, the rain was pouring
in volumes which, as it seemed to me, threatened to wash the town away.
Yet in an hour, the sky was bright and clear as ever. In another hour,
the streets were in splendid condition, but dusty as ever the next morning.

Mauna Loa is the mountain and Kilauea is the crater; the crater is,
so to speak, a boil on the flank of Mauna Loa and yet the crater itself
does not overflow. When the lava pressure becomes too great to be
restrained, it breaks out somewhere away down the mountainside. By
and by, the "rock sharps" say, it will burst out away offshore and then
we shall have an addition to Hawaii.

Another fact, and a very surprising fact indeed to me, was that the
nearer one gets to the volcano, the more heathenish the natives are.
With every mile's travel toward the volcano, I saw more signs of idolatry till, as we emerged from the forest on the rock flat, our native at-
tendants showed themselves regular Pele fanatics.

Even "Old Antony," the guide, recommended to us especially for
his Christian character, "backslid" when he saw the smoke and hinted
that it might be as well to gather some berries to offer as a sacrifice.
This comical retrogression in faith equally with progression in altitude
reminded me of the formula in use in Texas when I was there in 1867:
"There is no Sunday west of the Trinity, no law west of the Brazos,
and no God west of the Colorado."

Everybody knows how Pele, the devil goddess of the Kanakas, used to hold her fiery court in the flaming lake; how the filaments of lava, like glassblower's "thread" thrown off by the fiery waves were believed to be her hair; how the natives, when warned by well-known signs that an eruption was near, threw many fat hogs and other articles of value into the lake and, finally, how the brave princess Kauikulani, to convince her people that idolatry was foolish, descended into the crater without the usual ceremonies and returned unhurt to her amazed people. Is it not all written in mission reports and Sunday School books?

Brigham Young said the Kanakas are like the American Indians, descendants of the ancient Israelites who had "backslidden" and been cursed and turned dark accordingly. If so, they have slidden back a great deal further than they are ever likely to slide forward again. Some say that the topography is the cause of this local backsliding. Maybe so, but it is almost impossible to describe the topography. In truth, a very large part of Hawaii consists of a high tableland enclosed in a sort of triangle between the three great mountain peaks of Mauna Loa, Mauna Kea and Mauna Hualalai, and this plateau is a dreadful wilderness of tropical vines and giant trees, growing among and hiding immense rocks and crevices, with here and there a bare field of splintered lava or loose ashy stuff, and more rarely a fertile little valley.

Through all this century, while civilization was rapidly progressing and later, while Christianity was becoming dominant, the Kanakas were fading away at an unprecedented rate. Beyond the known causes—local wars, an unidentified pestilence, diseases introduced by profligate seamen, and leprosy—there are causes not understood. A too-rapid civilization is often fatal to primitive races. The sudden inflow of new impressions on unprepared nerve centers breaks them down.

The wearing of clothes seemed, for a time, to make the native women barren, as they paid no attention to the ordinary rules for keeping their raiment in healthful condition. Another cause has been lately assigned in these words: "Hawaii is a holiday land—a lazy land—a soft, luxurious, voluptuous land—and the depression from satiety is correspondingly great. There is no set purpose in life, no firm and high resolve, and so, when melancholy comes, it kills."

No other man dies so easily and gently as a true Kanaka. He can lie down and die whenever the notion takes him and with no apparent disease. This happens among many dark races, but with none so often as the Kanaka. In 1822, the missionaries placed their number at 142,000; by 1871, there were fewer than 60,000; as the pure Kanakas decreased, the foreigners and mixed breeds increased rapidly.

Honolulu is the Paris of the North Pacific. All the allurements of life abound. Hard work is left to the Chinese, for the most part, and active business to other foreigners. With at least half the population, social life is the only life. Receptions, teas and club meetings, political demonstrations, visits to the populace and to vessels in the harbor, rides and drives, poi suppers and bathing parties, chatting in the groves and lounging and smoking in the gardens, such is life in Honolulu.

If I should attempt to sum up on Hawaii-Nei (that is, the whole island group) in one sentence, I would say it is a land where everything seems to be something else. There is a riotous abundance of useful vegetation, yet soon or late, a blight comes on almost everything. There is a greater variety of fruits than in any other land I know, yet so far as I could learn, every fruit degenerates in a few years, and the stock must be renewed. The climate is almost perfection, yet measles, smallpox and affections of the heart are peculiarly fatal. The mountains look as if fixed in place for all eternity, and the more solid and rugged a mountain looks, the more certain it is to quake and more likely to split open. The natives are all well-educated and nearly all thoughtless. They are all nominal Christians, yet chastity is but a vague sentiment and many a common Kanaka will tell a lie when the truth would better serve his purpose.

Life and property are as safe as anywhere on earth, though the laws are not severe or executed with any special vigor. This may be due partly to the lack of acquisitiveness among the common Kanakas and their general live and let live feeling. They are the most unselfish race on earth. In a detached group in the interior, the property of any one is the property of another, if that other needs it badly.

A foreigner who has befriended one is the friend of all in the village. The men give him fish, fruits, flowers and poi; the women give what they have, and with a certain freedom which is delicate and charming. Unfortunately, their kindness may be a danger, for they may be affected with leprosy long before they learn the fact, and physicians have decided that, of all supposed ways, that is the one absolutely certain way of acquiring it.

The great product of the islands is the taro root, from which poi is made. It grows wherever the soil is kept thoroughly water soaked and in matted bunches very much like the calamus or sweet flag of the Middle Western states. The root is something like a long beet and, when boiled and mashed, the pulp is poi—a whitey yellow batter at first, later, a sort of sour paste. It operates homeopathically, so to speak —that is, it builds up a dilapidated constitution surprisingly, but if used exclusive of any other food, it injures the health.

There is no particular danger of any American or Englishman using the native style of it to excess, for it is emphatically nasty. Decently prepared, however, it is quite palatable. Local statistics say that one acre of it will furnish yearly bread for five thousand men—a pleasing statement which, I regret to say, I do not believe. Fish is said to be the proper corrective to take with poi and fish also are abundant.

They tell many hard stories about the former Hawaiians, that thousands of men were killed in the saturnalia following a king's death and thousands more were sacrificed or eaten, but as I saw nothing of the kind, I will continue to hope they are not true. There is peace and good fellowship enough now. There are social enjoyment and ease and joyousness and feasting enough now. Everywhere this is true, but especially in Honolulu.

There, the slant-eyed Chinaman and the darker Malay, the very fair Englishman, dark Portuguese and medium American, the smooth brown Kanaka and the rough brown Samoan jostle each other good humoredly in the streets and even practice a sort of social equality unthought of in the United States.

So whaling's embrace was only half-rude in Hawaii, as contrasted with its impact upon the people of the Arctic. The Sandwich Islanders were exploited aboard New Bedford whaleships but, on the other hand, they were paid *something* and for a very practical reason —it was desired to ship another crew of hard-working Kanakas next season.

The island natives did get venereal disease and a number of other things from the whalemen that they could well have done without and would not have gotten if left alone. Still, the whaling industry didn't jeopardize their survival, as it did that of the Arctic people; rather, it strengthened the islands' economy sufficiently so that the latter withstood successfully even the end of whaling, when that came.

Yet the embrace was at least half-rude, and no mistake. In part, it was because, as in New Bedford, the whaling business stimulated the accumulation of money so rapidly as to lead men into ways beyond bearing with perspective. In even larger part, it was because, even if the principals involved had spoken the same language, totally different temperaments (as with the Arctic people) would have prevented them from understanding each other.

It is a bald truth that the New Bedford whaleman, on the quarter-deck or in the forecastle, liked what he found in Hawaii, but he certainly didn't consider the Kanaka, or any to whom he applied the term, an equal. Instead, the Yankee looked upon the native as an agreeable, usually industrious child and did not ponder upon, or in fact even notice, that in the long run, he left this child far poorer in important ways than he found him.

# 12.

# Departure

With the opening of navigation in high latitudes, came increased perils. Not sufficient
were the dangers from their gigantic prey, or
furious gales, or the losing sight of the ships;
to these must be added the risk of being ground
between two mighty icebergs, of being caught
in some field of ice and forced ashore, of having the stout timbers of their vessel pierced by
the glittering spear of some stray berg as it was
driven by the force of polar currents. [Yet]
the season in [the] northern sea lasts but two
or three months and the temptation to incur
many risks for the sake of rapidly filling the
ship is too great to be withstood.
—STARBUCK,
*History of the American Whale Fishery*

Forty whaleships, a majority of them owned in New Bedford and
three of them, including the new *Concordia*, owned by George and
Matthew Howland, fitted out in Honolulu for the Arctic whaling
season of 1871 and there was not anything quite like the process
anywhere else—or ever thereafter even in Honolulu.

Still, there was no way of knowing that it was the last time. Matters went forward as they always had and as if they would continue
to do so forever. Such is the nature of man.

Whaleships sailed out of many ports, but never in concert, in such
numbers as from the Sandwich Islands. This fact arose, not from
sociability, but from practical accommodation to the nature of the
bowhead and the Arctic. The whalemen were all concerned with
making the best of the briefest of seasons, strictly delineated by the
going and coming of the ice. Nowhere else in the world did such a
large fleet congregate and operate in such close proximity in pursuit

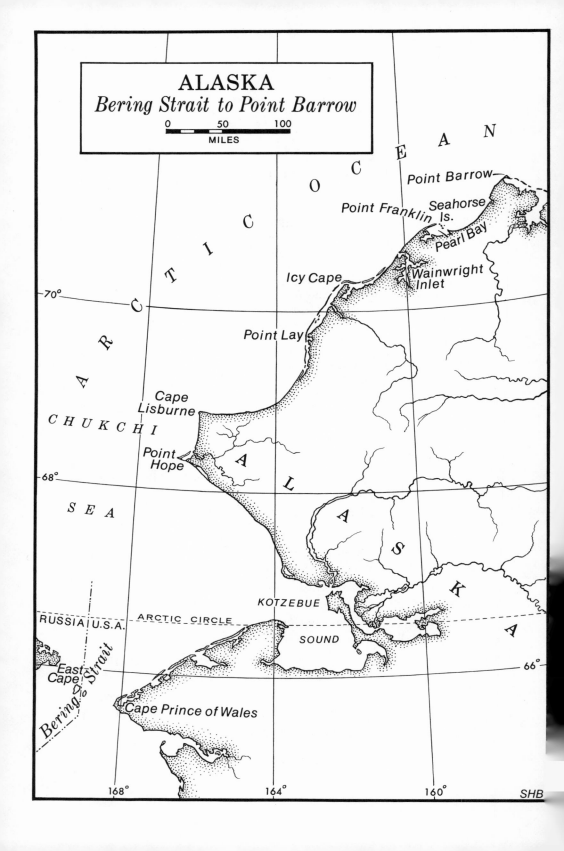

# ALASKA
## *Bering Strait to Point Barrow*

```
0        50        100
        MILES
```

O C E A N

Point Barrow

Point Franklin  Seahorse Is.

Pearl Bay

A R C T I C

Icy Cape  Wainwright Inlet

—70°

Point Lay

Cape Lisburne

C H U K C H I

A L A S K A

Point Hope

—68°

S E A

KOTZEBUE

ARCTIC CIRCLE

RUSSIA | U.S.A.

SOUND

East Cape

66°—

Bering Strait

Cape Prince of Wales

168°                164°                160°

SHB

of the whale. Although once through Bering Strait, the vessels might stand to either the west or east for the season, depending on hunch or inclination, they all had to be at about the same place at about the same time in order to have the best chance of achieving the same aim —a profitable voyage and a safe return.

Also unique to the bowheader who fitted out in the Sandwich Islands was the fact that he sailed from some of the best weather in the world, and tropical at that, into some of the worst, and polar at that. It was enough to ask of a New Englander, anatomically designed to stand an atmosphere that can produce 100 degrees Fahrenheit above zero and ten below in the same year, on the same corner of Boston's Boylston Street. To ask it of a South Sea Islander seems, in retrospect, unspeakable; he was especially vulnerable to respiratory ailments and Arctic weather took its toll of Kanakas forced to voyage from palm trees to icebergs in roughly four weeks' time, a harsh transition almost faster than the mood and spirit, let alone the body, can keep up with.

There is nothing comparable to the departure of a fleet. Even a landsman finds its related activity exciting and its accomplishment so final as to be saddening even if you know no one aboard. Once the seemingly interminable buyings and bustlings, the back and forth between ship and shore, the fitting and financing, the arrangings, hirings, signings and designings of preparation are over—and this keeps the harbor in healthy turmoil for days—there comes the time to leave and, in this case, the time is spring.

Leaving is many things to many men. The whaling captains, a few with their families aboard, most with not; a few with homes in the islands, most with not, were ending a pleasant social interlude. During their stay in the islands, the masters convened regularly, afloat or ashore, yarned, smoked and ate together, compared professional notes with such instinctive caution that no one revealed anything of real value yet all appeared genial and forthright, and this rare opportunity for commingling with one's peers was an agreeable change. Now, each returned to the comparative isolation of his own quarterdeck.

The Kanaka was leaving home and family, but only for a short time. Besides, the employment, and the money, however little it might be, looked good to him. More and more, reflecting the increasing dilution of his culture as it responded to the foreigners' commercial activity, he wanted what only money could purchase.

By contrast, the white man in the forecastle had long since—at least, months and perhaps years ago—said goodbye to whoever or whatever meant something to him and on any given day had no way of knowing whether the person or thing still existed. One of the most awful instruments of suspense is a four-month-old letter that has beaten its way around the globe which begins, "I am not feeling very well . . ."

Far too little ever was recorded, and little of this ever preserved of what these men thought, of what feelings preoccupied them. Neither time, space, opportunity nor surroundings encouraged the keeping of a journal. Especially, Arctic whaling did not, for it was primarily a matter of haste, the whales were transient, the weather was imperious. Moreover, many of the men in the forecastle could not write at all and those who could wrote cumbersomely; they were laborers called upon for certain special skills (such as not getting seasick and being able to go aloft without falling) engaged in slaughterhouse and rendering occupations that have been much romanticized, principally by those who never went whaling—but laborers still. It is hard to hold a pen in a fist the size of a ham.

Above all, what they felt must have involved those matters most difficult for any writer, even a good one, to express—loneliness and homesickness, the dried-up hopelessness born of a half-dozen kinds of unfulfilled wants and needs so obviously about to remain unfulfilled that to dream of what it might be like if they were otherwise was no more than a cruel self-indulgence.

I have seen penciled fragments of what some did write about these things on the pages of battered, yellowing copybooks. As a rule, these records were started early in the voyage, when the memory of what had been remained bright and sharp, the taste of the kiss, the sweet smell of the drying hay, the etching of the roof peak against the night sky. They started with bold flourishes of signatures on the inside cover: "Jonathan Abbott, His Book," and all letters largely done and with the author's ego showing.

The first entries were lengthy, interminable sentences of crabbedly contrived misspellings about the weather and whales or no whales, and references, sometimes with a delightfully unconscious humor or poetry, to the largely dull task with which a sailor aboard a whaleship filled his days.

Soon, the entries became shorter, the bare bones of larger thoughts, and you have to force yourself to think what the author is not saying in order to realize what he *is* saying when he speaks, in a single sentence nevertheless repeated for several days of "wether is retched," or constantly aching teeth, or constipation, or stomach disorder that will not let him sleep, or childlike references to "my deer wife," in which he attempts to relieve his agonies without even the knowledge that use of an exclamation mark would better reveal their terrible intensity.

The long paragraphs become short paragraphs, and finally, there are no paragraphs at all, and by the time you get to page ten of the copybook (some did not even survive the first nine pages), there is nothing but a large water spot on the century-old paper. If you are romantic, this was a tear; if you are not, the deck leaked over the man's bunk and it is salt water. In any event, it was too hard to write, and there was too much to write, and so one soon wrote nothing at all because writing means thinking and it is better to keep busy with the comforting simplicity of mallet and twine, tar and needle than to dwell upon aching teeth, protesting stomach and "deer wife."

So sometimes, perhaps even many or most times, the Yankee whaleman in the forecastle faced the morning of departure retching from the rum of the night before or wondering uneasily whether he had caught anything from that woman the night before. In any event, he seldom had much to leave in foreign ports—unless he was of the exceptional minority soon bound aft for an officer's berth because of his aptitudes. The average sailor ashore did not travel far from the waterfront; the ship, good or bad, was his only sanctuary and held his only friends of the moment. He was not inclined

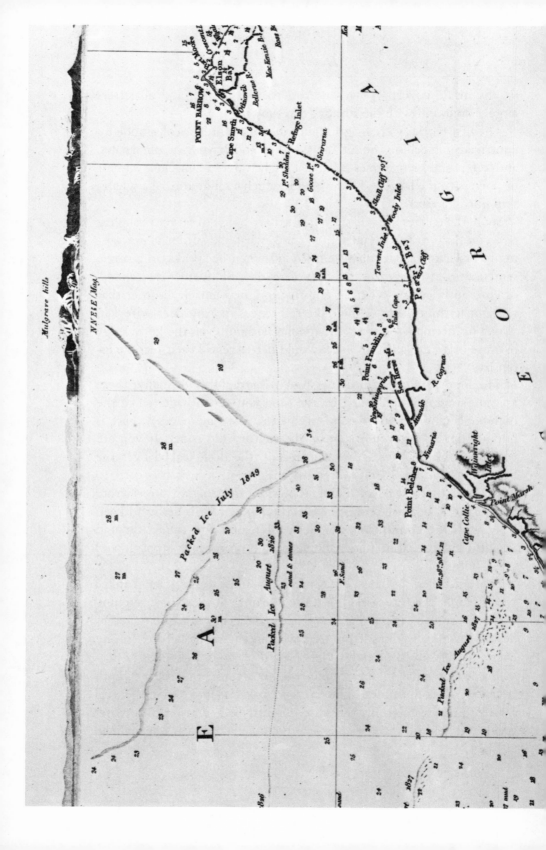

The Northwest Coast, from Point Rodney to Point Barrow. (In the absence of adequate U.S. surveys of Arctic waters, British or Russian charts of the northwest coast were in general use by the New Bedford fleet until late in the 19th century. *Courtesy of The Whaling Museum, New Bedford, Mass.*

to do the sightseeing that his unique job made occasionally possible; instead, he went ashore to satisfy his frustrated appetites. Since the vendors of sex, liquor, food, and inexpensive native craft articles stormed down to the highwater mark to meet him, he never did get to see much of any port, and thus considered all ports essentially alike. Besides, he was no more at home with landsmen abroad than he was with those back in New Bedford; the simplest communication is impossible, even granted a common language, when farmers and sailors spice conversation with the idioms of their respective callings. What plowman, either here or there, is going to understand the ultimate empty-headedness, the frustrating and even dangerous uselessness of a man aboard ship who does not know the difference between Charlie Chace (a high-line whaling master of the New Bedford Wing-owned fleet) and Charlie Noble (the smokestack of the ship's galley)? And what sailor is going to try to explain it to him? It's not worth the bother.

In any event, it is well-known that work will take care of the morning-after queasiness. The time has come to get underway; soon, the sun will be off the horizon and half the day gone. All hands are rousted out, the windlass drum rolls ponderously, until the anchor chain is up and down; the Old Man is on the quarterdeck, with an eye for everything simultaneously.

For days, the fleet has presented only the tracery of rigging and empty spars against the sky; now the weathered canvas sprouts everywhere, like instant leaves miraculously clothing long-bare branches. The blocks chuckle and squeak, depending on wear and lubrication; there is some sound of the authoritative voice upon the still air; the small black figures of men crawl out along the yards; the great iron links clank through the hawse pipe as the dripping anchor breaks the surface. We are at the moment now when a stationary combination of humans and objects, often awkward and ill-married in superficial appearance, let loose their last ties to the muddy earth crust and, in no time at all, become the enigmatic and compelling entity that is a ship.

Fore-topmast staysail, jib, and flying jib fill, as the vessel's bow falls off slowly from the wind's eye; foresail and main belly in the breeze. The man at the wheel brings up the helm now, anticipating the course the vessel will settle on when the last flutter of canvas ends. He gives her a good rap full, and the water begins to bubble around her curves. How ingeniously contrived is this New Bedford instrument, something of a windmill in motion, and since its blades of sails, slantwise to the breeze, cannot turn, something else must move and thus, ship and all, corned beef, cabbage, and casks, harpoons and harpooners are thrust irresistibly toward the Arctic.

There is nothing like it. Forty ships, with fifteen hundred souls aboard, including a handful of women and children, a piece of the world suddenly gone traveling, as if New Bedford, or Dartmouth, or Mattapoisett, or any other place, suddenly broke away from its continent, together with all of its goods, chattels, and populace, its Bibles, bedpans, accents, and aspirations, and moved about, who knows where, until it decided to come back once more to where the chart said it was.

This harbor, yesterday populated with a community, and all of the latter's dreams, animosities, frustrations, dark thoughts and laughter, now lies empty; profoundly empty, without sound or substance, except such as nature, rather than man, provides. One great gull seesaws over the anchorage in which nothing is now anchored, probably wondering in the silence of his soaring where everybody went all of a sudden, and already missing, without thinking about it, the daily garbage thrown overboard from the galley.

There they go—great square sterns, sharply rounded sterns, sterns with tumble home, *Concordia*'s stern, with the carved eagle; all commanded by their massive rudders that urge them northward. The fleet settles lower and lower upon the earth's curve until finally the topsails and even the topgallant sails drop beyond seeing and the silence washes back upon the shore where men walked only yesterday and walk no more.

Some fleets may be described as proud; this one is practical. The romanticist may think of its masters as knights off on a crusade to joust with the Arctic. It is more reasonable to think of them as engineers and production managers and to ignore the fact that a ship under sail, any ship, persists in inspiring romantic notions about its calling and its people.

If you want to know how very practical they were, consider the following (obviously prepared by the owner or agent), taken from the records of an Arctic-bound Yankee whaleship sailing from Honolulu:

"Instructions to be perused frequently during the voyage by the captain and mates:

"You will see that the blubber is properly tried, that the oil does not get burnt or discolored in boiling, and that the blubber does not become rancid before boiling. Either injury is equally detrimental to its value and greatly injures its sale.

"You will see that the bone be properly scraped and cured and neatly bundled before arrival. It will be best to leave it open to dry as long as it will answer before putting it up and stowing away.

"You are not to omit taking a whale when you can, even if you are obliged to suspend boiling or other duty.

"You will avoid visiting other ships or receiving company on board of your own to a prejudicial extent.

"You will exercise prudence in all your expenditures."

Yes, indeed, very practical.

Still, it is only fair to mention that at least three masters in this fleet—Aaron Dean of the bark *John Wells,* Timothy C. Packard of the *Henry Taber,* and Captain Leander C. Owen of the *Contest*—had a mission that came ahead of their whaling. They were pledged to search for survivors, if any, of the whaling bark *Japan,* lost in the Arctic the season before.

# 13.

# The Wreck of the *Japan*

July 9, 1871—At noon, Linus Kenworthy, a
native of London, England, died with scurvy.
We took this man from Icy Cape. He was one
of the unfortunates of the late ship *Japan*.
July 10, 1871—This morning, took the corpse
ashore and buried the poor man, halfway between Harrington's Inlet and Point Belcher.
—Log of the bark *Henry Taber*

The loss of the *Japan*, commanded by a New Bedford native, Captain Frederick A. Barker, and the rescue of those of her crew who
survived by New Bedford whalemen in the spring of 1871 provided
fresh evidence, if any was needed, of what Arctic whaling was about.

Because Captain Barker and his men had to winter with the Eskimos and because he was sensitive and articulate, New Bedford whaleship owners were once more reminded that an everlasting triangle
of hostile components was involved in chasing the bowhead: whaleships versus whales; weather versus whaleships, and the whaling industry versus the Eskimo. The smell of death hung about all three
sides of the triangle. Sensibility recommended at least its suspension,
yet it endured and was fiercely, recklessly perpetuated, even over the
misgivings of some who knew it well, because in the perpetuation,
there was money for the larger and stronger of the cultures involved.

During the latter end of the whaling season of 1870, the weather
had been so good and whales so plentiful that Barker had put off
his return through the Bering Strait until rather late. He was reminded of this because the cold was becoming intense and ice had
formed along the coast, rendering navigation less pleasant and more
difficult each day.

Thus, he began working the ship to the southwest toward the

strait, when thick weather set in. On the second of October, the bark was off Cape Lisburne, near Rendezvous Bay, latitude 68, and Barker stood to the northward to give the headlands and currents a wide berth while passing into the Kamchatka Sea. Two days later, no ships were in sight and it began to blow, developing into a heavy gale in which the vessel labored for more than four days. The low temperatures persisted, the sea was very heavy, and the *Japan*, even under shortened sail, buffeted and strained sufficiently so that the captain knew she was in trouble.

On the eighth of October, a full day before the wind slackened, they sighted the whaleship *Massachusetts*, Captain West Mitchell, of New Bedford, and made signals for her to keep company with them, because, Barker had concluded, "I knew full well that we could not weather the storm and feared that our end would come before other assistance could reach us." Probably those aboard the *Massachusetts* never saw the signals. The visibility remained very poor.

On the morning of October nine, although it was impossible to determine the ship's position, the *Japan* was off East Cape; this is just west of the Diomedes Islands, and is the extreme tip of the continent of Asia.

In Barker's words, this is what happened:

The gale blew harder, if possible, attended by such blinding snow that we could not see half a ship's length and were obliged to light our binnacle and cabin lamps. Near noon, the weather still continuing very thick, we discovered breakers on the lee bow, close aboard.

The ship had been running under lower topsails and stormsails, but owing to the strength of the gale, was making racehorse speed. The helm was put starboard and braced up sharp, when we discovered breakers off the weather bow, close aboard. Just then, to add to our horror, a huge wave swept over the ship, taking off all the boats and sweeping the decks clean.

Our situation was now most critical, death truly stared us in the face. But one chance remained for our lives and that was to run the ship's bows on a few yards from where the sea struck us, which I did. Another sea struck us and sent us broadside on the beach. The beach was rocky and steep, but all the crew managed to reach dry land in safety.

I remained below decks for three hours searching for clothing, provisions and spirits for the relief of my men and hoping that the vessel would meanwhile be washed higher up on the beach. I watched my men struggling through the surf, attempting to secure some things, and I finally swam ashore and the natives met me and put me in one of their sledges and conveyed me to their settlement.

Some distance from the beach, we passed on the way the bodies of sailors frozen to death. The air was piercing cold and several of my men, being unable to dry their clothes, had fallen by the wayside and died. I discovered that out of a crew of thirty, already eight had frozen to death. All were badly frostbitten.

When the natives dragged me out of the breakers, I was almost breathless and nearly frozen. Once they placed me upon a dogsled and started for their huts, I thought my teeth would freeze off and I could not endure riding any longer. I got off the sled and started to run, but fell as one paralyzed. The natives again placed me upon the sled and held me on and seeing I only obstructed their movements by my attempts to walk, I kept still, as indeed I could not do otherwise, being completely helpless.

I suffered most dreadfully from the cold and supposed I was freezing to death. In a short time, we reached the huts and I was carried in like a clod of earth, as I could not move hand or foot. The chief's wife, in whose hut I was, pulled off my boots and stockings and placed my frozen feet against her naked bosom to restore warmth and animation.

I say it with the deepest gratitude and thankfulness that only for the kindness of the natives, every soul, myself included, would have perished on the beach as they landed, as there was no means at hand of kindling a fire or of helping ourselves one way or the other. The natives took the whole crew in their huts, as they had done by me, and kept them from freezing to death.

Since the whaling fleet of '70 had all gone south by this time, the survivors of the *Japan* shipwreck were forced to winter with the Eskimos, whose kindness, nursing, and hospitality kept them alive— but it was still a hard time for the white men.

There was, first of all, the matter of food. The few casks of bread and flour that were washed ashore from the whaleship did not last very long and then it was necessary for the crew to live as the natives did, on raw walrus meat and blubber, and neither of these was very sweet or fresh most of the time. The whalemen did not take to the diet readily but, as always, the desire to live was strong. Barker starved himself for nearly three days before he could bear to try the Eskimos' food but finally "saw that I must do it or die. Hunger at last compelled me and, strange as it may appear, it tasted good to me and before I had been there many weeks, I could eat as much raw meat and blubber as anyone, the natives excepted, who are enormous eaters."

It became immediately obvious to Barker and his men that, because of the activities of the whaling fleet, the Eskimos did not, in fact, have enough to eat. Barker soon learned to speak the native language while waiting out the winter at Cape East and thus came to understand the Eskimo and his problems far better than did most of his fellow whalemen. He concluded, "Should I ever come to the Arctic Ocean to cruise again, I will never catch another walrus, for these poor people along the coast have nothing else to live upon.

"I felt like a guilty culprit while eating their food with them, that I have been taking the bread out of their mouths, yet although they knew that the whaleships are doing this, still they were ready to share all they had with us."

After the *John Wells, Henry Taber,* and *Contest* had stopped at every place where survivors of the *Japan* were living, including Owalin, the site of the wreck, had picked up all the white men and made contributions of equipment and provisions to the Eskimos in appreciation, one of Barker's first acts was to draft an appeal in behalf of his native benefactors, for publication in the *Whalemen's Shipping List and Merchants Transcript.* He wrote:

I wish to say to the ship agents and owners in New Bedford and elsewhere that the wholesale butchery of the walrus pursued by nearly all their ships during the early part of each season will surely end in the extermination of this race of natives who rely upon these animals alone for their winter's supply of food. If this is continued much longer, their fate is inevitable, as already this cruel persecution has been felt along the entire coast, while a wail like that of the Egyptians goes through the length and breadth of the land. There is a famine and relief comes not.

I have often been asked by these simple, kind-hearted children of the North the reason the white men that came in ships were so cruel as to take away their food and leave them to starve, that if persisted in, contrary to their wishes and remonstrances, they would not provide for the castaway thrown upon their shores, as they could not rob their children of food to feed the strangers.

They said that when the ships first came, they were pleased, but now they all caught the walrus and they were hungry everywhere, and there was no more rejoicing at their appearance. The capture of the whale did not affect them much, as they were not their dependence, but the walrus were, and would I not intercede with the President in their behalf on my return?

I can but appeal for favor from the hands of those from whom they have suffered most, that they may have their attention drawn to this important subject (of life interest to them) and prohibit their captains from carrying on this war of extermination.

I have conversed with many intelligent shipmasters upon this subject since I have seen it in its true light and *all* have expressed their honest conviction that it was wrong, cruel and heartless and the sure death of this inoffensive race, that they would be only too glad to abandon the thing at once, if their employers would justify them, but until the subject was introduced to public notice, they were powerless to act, for fear of not being sustained by them.

For one, I will never trouble them more, as their flesh has been the principal subsistence of myself and men during the long months of our weary imprisonment that alone enabled us to survive its fatigues.

To abandon an enterprise that, in one season alone, yielded 10,000 barrels of oil, for the sake of the Esquimaux, who have found an advocate in one who has passed a few months with them, may seem preposterous and meet with derision and contempt, but let those who deride it see the misery entailed throughout the country by this unjust wrong, with death knocking at the door, while hunger was staring through the window, as I have, during my travels, and I feel quite sure that a business that can last not longer than two or three years more will be condemned by every prompting of humanity that ever actuated the heart of a Christian.

That Captain Barker was not alone in his concern for the plight of the Eskimo became immediately obvious.

The Hawaiian *Gazette* published a letter to the editor signed "Whaleman" which related:

The wholesale killing of walrus in the Arctic regions endangers the starving Eskimo and many whalemen have expressed a determination on that account to desist entirely from slaughtering them.

Years ago, whales were so tame there that the Eskimos had no difficulty in killing them as often as they were required for food. As the whalers became plenty, the whales became correspondingly shy and the Eskimos had to depend on walrus and seal.

For four or five years, the whalemen have been catching walrus and they are now few and shy and the natives are on the verge of starvation. This season [1871], some of them were seen to have nothing but tough walrus hide to eat and some of them have starved to death.

A writer signing himself "Shipmaster," who addressed a letter to the New Bedford *Republican Standard*, noted:

For the past three or four years, (1868–71) the North Pacific whaling fleet have been taking walrus in the months of July and August, as the whales in those months go into the ice and around Point Barrow, out of reach of the whalemen.

During all the years from 1849 to 1867, the whalemen had left the walrus alone or taken a very few. In 1868, a few ships commenced taking walrus and did quite well, securing from two hundred to six hundred walrus and destroying half as many more.

In 1869, a large number of ships were engaged in the business; in 1870, the whole fleet, with two or three exceptions, went in and took all they could. Probably not less than 50,000 female walrus, with their young, were killed and destroyed. The past year, three fourths of the fleet were engaged in the business but the walrus were shy and far into the ice and they did not do as well. Shipmasters had to send their boats twenty and twenty-five miles [into the ice] to find them.

The Arctic walrus are nearly all females who go into the Arctic in the summer months to bring forth and nurse their young, which the mothers are very fond of and attached to. They will never forsake their young, but will take them in their flippers and hold them to their breasts, even when their destroyers are putting their sharp lances through and through them and the blood streaming from every side, uttering the most heartrending and piteous cries and so until they die, and then the little one must starve unless the whaleman can thrust his lance through it and send it to the bottom.

This is one of the most cruel occupations that I know of and many a humane whaleman has felt guilty and turned aside as he did it. The walrus average about twenty gallons of oil and four pounds of ivory.

But the worst feature of the business is that the natives of the entire Arctic shores, from Cape Thaddeus and the Anadir Sea to the farthest point north, a shoreline of more than one thousand miles on the west coast, with the large island of St. Lawrence, the smaller ones of Diomede and King's Island, all thickly inhabited, and our own coast of North Alaska, are now almost entirely dependent on the walrus for their food, clothing, boots and dwellings. Twenty years ago, whales were plenty and easily caught, but the whales have been destroyed and driven north, so that now, the natives seldom get a whale. This is a sad state of things for them.

The question now is, should our whalemen keep on taking the walrus and eventually starve and depopulate these Arctic shores? It will certainly come to that soon; already, they are starving or on the point of starvation. Several captains lately arrived home have told me that they saw the natives thirty and forty miles from land on the ice, trying to catch a walrus to eat and were living on the carcasses of those that the whalemen had killed. What must the poor creatures do this cold winter, with no whale or walrus?

Captain Barker, who was shipwrecked and passed the winter with them last year, says they were upon the point of starvation in many places on account of the walrus being so scarce and shy and he was ashamed of himself to think that he had been engaged in the business and would never do it again. I have seen most of the captains lately arrived home and they all tell the same story, that the natives are starving, or will starve, if the business is not stopped.

Some say, "I will never take another walrus," but several others I have talked with say they won't take walrus if others will not, which means just this: I shall take all I can. But it wants the condemnation of the shipowners and agents here in New Bedford, for I think their ships can be better and more profitably employed in whaling. There are plenty of humpback and California gray whales yet south of the Arctic and long sperm whale cruises will pay better than the early bowhead whaling. Ships can engage in sperm whaling until June or the first of July; ships the past five years have not more than paid their ice damages up to the first of July.

I think this is the opinion of most shipmasters; at any rate, I ask my brother shipmasters to spare the walrus and let the hospitable, kind and good natives of the Arctic shores live.

"Man's inhumanity to man makes countless thousands mourn," says one, and now, this cold winter, I have no doubt there is mourning in many an Arctic home as the little ones cry for something to eat and the parents have nothing to give, for the walrus are killed or driven far away.

There is no evidence that the practice of killing walruses was, in fact, "condemned by every prompting of humanity that ever actuated the heart of a Christian," even though the New Bedford whaling industry was dominated to a singular degree by active Christian leaders, including George and Matthew Howland. At the end of 1871, in its roundup of the year's whaling activities, the *Republican Standard* noted there was sentiment in behalf of sparing the walrus for the sake of the Eskimo but it devoted only a couple of lines to the matter and made no editorial comment. And as a matter of record, New Bedford whaleships continued to kill walruses in large numbers.

# IV

# The Loss of the Fleet

# 14.

# Through "Seventy-two Pass"

In the second half of the nineteenth century,
when the great bowhead whales were getting
ever scarcer, the whalers hunted anything that
would help to make their voyage profitable.
—*The Whale*, TRE TRYCKARE

Even as the prospects of quick gain made Arctic whaling unique, so did its extraordinary demands upon the shipmasters, and it was no accident that most of the captains of the fleet of 1871 were experienced bowheaders of the highest competence, including Captains Tripp of the *Arctic;* William H. Kelley, *Gay Head;* Hezekiah Allen, *Minerva;* George W. Bliven of the *Elizabeth Swift;* B. F. Loveland, *Reindeer;* Daniel B. Nye, of the *Eugenia;* Henry Pease of the *Champion*, and Robert Jones of the *Concordia*.

It is significant that Captain Jones was given command of the prize of the Howland fleet and that although, in Matthew Howland's letter book, he makes pointed criticism of other masters' costs and accomplishments, it is to Jones that he wrote—and Jones alone— "[Brother George and I] hope and trust your success will continue."

From Honolulu to the Howland wharf in New Bedford, these men, and virtually all the other masters of the bowhead fleet, were known both personally and professionally; anybody of consequence in the industry could rate these captains, without reference to documents, on the basis of what their catch had been in recent voyages. Perhaps the closest parallel today is found in the batting averages of major league ball players. There were those mates and masters who argued, with certain reason, that an officer should not be penalized—often by loss of command—for one or two poor trips that might well have been caused by circumstance rather than lack of skill or poor judgment—but he usually was, nevertheless.

So the master was not expected to make mistakes, of any kind. It was hard not to make mistakes in Arctic whaling.

In the beginning, the fleet breached the Aleutian chain and entered the Arctic Sea. Often, this entailed a beat to windward through the slot at Amukta, just west of the Islands of the Four Mountains, which the whalemen called "Seventy-two Pass," because this channel linking the North Pacific and the Bering Sea lies on the 172nd meridian of longitude.

From that moment on, the master faced extraordinary, and quickly changing, problems of navigation and shiphandling.

Aldrich, who wrote extensively on nineteenth-century Arctic whaling, has commented that "ice navigation requires consummate skill . . . it is a peculiar kind of navigation. A contrary current may hold you in the pack while others may make sail all about you. You may be within easy sailing distance of a passage through the ice and not know it. There is always the danger of being stove; it requires infinite patience, and it is complicated by rain, fog and gales."

He related an incident in a matter of litigation, in which an Arctic whaling master was asked what he would do if he found himself "on a lee shore [that is, with the wind blowing toward the land] in a gale, where it was impossible to tack ship, where there was not room to wear ship [in other words, there was no room to turn around and head offshore], and it was not possible to anchor."

The landsman who asked the question obviously expected the captain to reply that he would expect the ship to go ashore, but the latter did not. Instead, the master said, "I would take in the after sails, haul everything hard aback and boxhaul her."

What this means in shore terms is that he would put his engine-less, square-rigged ship into reverse and sail her out of the dilemma backwards. This is a reasonably delicate maneuver because care must be taken not to damage the rudder, which is more exposed to the ice than usual, and because the ship's masts are stayed from aft—

that is, the principal strain upon them from wind blowing into the sails is not expected to come from ahead—and the risk of dismasting the vessel is always present in boxhauling. On the other hand, because many whaleships were bluff-bowed and narrow-sterned to increase stowage capacity, some may even have sailed faster backward than they did forward.

Sailing in and around ice chunks that ranged from six to sixteen feet most of the time, and complicated by "young" ice that was from two to four inches thick required the ship handler to maintain not only constant vigilance, but a drumfire of steady commands to the man at the wheel for hours on end. It was a matter of perpetual maneuvering when the obstacles themselves were perpetually in motion, of seeking openings, of trying to find other than dead-end channels, of dodging largely submerged dangers that weighed tons and could cause major damage to the ship's bottom in half a minute, of endeavoring to keep the vessel in water enough to float her, without benefit of adequate charts or navigational aids.

According to Captain Kelley:

One of the perplexities of the navigator cruising in the Arctic Ocean is the singular effect northerly and southerly winds seem to have upon the mariner's compass. Captains have noticed this singularity for years and no solution of the matter, as far as I have learned, has yet been arrived at.

Navigators have noticed that, with a north or northeast wind, they can tack in eight points, while with the wind south or southwest, in from fourteen to sixteen points. All navigators know that for a square-rigged vessel to lie within four points of the wind is an utter impossibility, the average with square-rigged vessels being six points.

This peculiar action of the compass renders the navigation of the Arctic difficult and, at times, dangerous, especially in thick, foggy weather. Navigators in these regions have proved to their satisfaction that on the American coast, north and east of Point Barrow, to steer a land course by the compass and allow the variations given by the chart of 44 degrees, 15 minutes east, with the wind at north or northeast, would run the ship ashore, steering either east or west.

Experience, therefore, has obliged navigators to ignore the variations marked by the charts and to lay the ship's course by the compass alone to make a land course safe in thick weather. With an east or west wind, the effect on the compass is not so great as with other winds.

I have said this much to show the workings of the compass in the Arctic Ocean during different winds, not that I admit that the wind has any effect whatever upon the compass. I give the facts as they came under my observation and corroborative testimony will be borne by any shipmaster who has cruised in the Arctic Ocean.

Weather was another persistent problem. This is Captain Pease of the *Champion* talking and he is referring to the bowhead season of 1870, the year before:

On the evening of October 7, it blew almost a hurricane; I worked the ship to south of Point Hope, with the main topsail furled; we lost the starboard bow boat, with the davits. The ship was covered with ice and oil.

On the 10th, we entered the straits in a heavy gale. When we were about eight miles south of the Diomedes, we had to heave to under bare poles. It was blowing furiously, with the heaviest sea I ever saw, and the ship was making bad weather of it.

We had about 125 barrels of oil on deck and all our fresh water. Our blubber was between decks [they had taken four whales in the last few days and bad weather had prevented them from trying out the blubber] in horse pieces and going from the forecastle to the mainmast every time she pitched, and it was impossible to stop it.

The ship was covered with ice and oil; we could muster only four men in a watch; the deck was flooded with water all the time; there was no fire to cook with or to warm by, all of which made it the most anxious and miserable time I have ever experienced in all my sea service.

During the night, we shipped a heavy sea, which took off the bow and waist boats, davits, slide-boards [on the sides of the ship, on which the boats slid when being hoisted or lowered] and everything attached, staving about twenty barrels of oil.

At daylight on the second day, we found ourselves in seventeen fathoms of water and about six miles from the center cape of St. Lawrence Island. Fortunately, the gale moderated a little so that we got two close-reefed topsails and reefed courses on her and by sundown were clear of the west end of the island.

Had it not moderated as soon as it did, we should, by 10 A.M., have been shaking hands with our departed friends.

About the first of May, the whaleships of the fleet of '71 began to arrive at the ice south of Cape Thaddeus. They found plenty of ice, so closely packed that the vessels made little headway to the north. The wind blew hard from the southeast most of the month, but about the first of June, the ice opened enough to let the ships up to within sight of Cape Navarin. Here, five or six whales were taken and a good many more were heard spouting in the heavy ice beyond reach, but they soon left.

The fore part of June brought light and variable winds, with a good deal of fog; by the middle of the month, the ice opened again and the fleet pushed to the north. A few whales were picked up during the crossing of Anadir Bay, but by the time the ships got to Cape Bering, and Plover Bay, the whales had all passed through the straits.

Here they sustained an early ice casualty. The bark *Oriole* was badly stove and put into Plover Bay where it was found impossible to repair her damage. The damage must have been substantial because the ingenuity of the bowheader in making repairs—and continuing to hunt whales thereafter—is a matter of record; in one instance, the master twice ran his ship ashore in a sheltered cove, hove her down to expose the lower bottom, by running a line from the maintop to an anchor ashore and hauling on it, and made a work raft for the men caulking and refastening by removing and lashing together several of his upper spars, which were replaced after the job was completed successfully.

Captain Benjamin Dexter of the *Emily Morgan* bought the *Oriole*, as was, for $1,350, stripped all usable and removable items from the beached wreck, and sold them to other ships of the fleet for a total of $2,541.17, realizing a net profit from the transaction of $1,191.17. Presumably, the crew of the *Oriole* was distributed among the other vessels, although the *Morgan*'s log does not mention this.

The fleet passed through Bering Strait between the eighteenth and thirtieth of June, various of them taking on board the survivors of the *Japan*. Not seeing any whales and finding large quantities of ice, the whole fleet now engaged in catching walrus. These were very shy and scarce, in comparison to former seasons, the boats frequently going fifteen or twenty miles into the ice to get them. There was much fog in June and July while they were walrusing and also large bodies of ice, the east shore being unapproachable until the very last of July.

The *Contest*, for example, commanded by Owen, passed Indian Point on the sixth of June, took aboard two officers and seven men of the *Japan*'s crew, found the strait much blocked with ice, but managed to take two bowheads, of a large number seen, which made about a hundred barrels of oil apiece. Drifting past East Cape on the sixteenth, they found walrus and killed four hundred in about a twenty-three day period, which made three hundred barrels.

On the night of July 17, the *Contest* got into the ice in the fog and, in working her way out, sustained bow damage at the waterline. But it proved to be repairable, and she was working her way up past Icy Cape and Blossom Shoals by the sixth of August, in company with eight or ten ships. She saw bowheads on the seventh, off Wainwright Inlet, and had taken five up to the twenty-eighth of the month. On that day, she got aground while underway to take a whale, but was assisted in getting afloat again by Captains James H. Fisher of the *Oliver Crocker* and West Mitchell of the *Massachusetts*.

In the latter part of July, the fleet experienced strong winds from the southeast and northeast, which broke up the walrus catching, and they pushed to the northeast for Icy Cape. The ice began to disappear from the east shore south of Cape Lisburne and the vessels moved to the eastward, the main body of ice being in about latitude 69 degrees, 10 minutes. They followed the ice in to the east shore, where they found a clear strip of water running to the northeast along the land.

In this clear water, they worked up to within a few miles of Icy Cape and some of them anchored, not being able to proceed any farther because of the ice lying on Blossom Shoals. At this time, the wind was blowing strongly from the northeast, a condition that lasted for several days.

On the sixth of August, the wind moderated and the ice started off the shoals. Several ships got under way and passed the shoals and, in a few days, most of the fleet was north of Blossom Shoals. The weather was much improved and they worked to the northeast as far as Wainwright Inlet. Here, they found whales and a number were taken at once. The ice was very heavy and closely packed, however, and a great many were lost. Still, the prospect looked favorable and hopes were entertained of making a large season's catch. All the ships were now either anchored or made fast to the heavy ground ice with ice anchors. Whaling was carried on briskly for several days, the boats cruising in the open ice.

Five days later, a large number of whaleboats were caught in the ice by the wind shifting and sending the ice in toward the shore. The wind was from the west and the ships were obliged to get under way to keep from being jammed; they worked inshore under the lee of the ground ice. With considerable difficulty, they succeeded in saving their boats by hauling them a long distance over the ice, an arduous task because of its unevenness. Some of the boats were badly stove in, but they were finally all recovered.

Captain Kelley said, "The ice closed up suddenly and we were forced to drag twenty-six boats over it. Fourteen boats were collected on a single cake at one time.

"Within half an hour from the time the ice began to move, we were solidly enclosed."

The ice kept settling on shore steadily, the ships kept fleeting into shoal water to avoid being stove. Some of them grounded, but were got off again without great difficulty. On the thirteenth of August, the ice halted its march, leaving an open strip of water along the land as far as Point Belcher. On that day, there were twenty ships in the area, some of them beset in the ice, others riding at anchor, and by the sixteenth, thirteen more had arrived. Boats were kept off whaling every day and the crews saw and heard plenty of whales among the heavy ice but could not get to them.

By the seventeenth of August, when all the fleet was lying on the outside edge of the shoals extending from near Wainwright Inlet to Point Belcher, a further movement of the ice toward them forced the ships to haul up anchor and get nearer the shore. Kelley, determined to get into the deep water inside the shoals—hoping that the oncoming ice would ground on the shoals short of the fleet—took a sounding lead and picked out a channel across them, marking it by sinking bundles of bricks to which were attached pieces of cordwood to serve as buoys. The other ships followed his lead and they all got inside, although a couple of vessels went aground briefly during the maneuver.

The fleet was getting some whales, but getting them the hard way because of inability to move the ships. They saw and heard far more whales than they were able to strike; the bowheads knew they were safe in the ice and they stayed there. Several whales that were struck either ran for the ice and thus parted the line and disappeared, or got into the ice after being hit hard and died there—within sight, but not within reach. "Struck a whale five miles from the ship," the mate of the *Henry Taber* noted gloomily. "Whale ran under the ice. Lost line. Shut down thick fog. Saw him no more." And again, he wrote, "Still fast in close-packed ice. Foggy part of the time. Saw three ships take whales today and we can't help ourselves. Leave one whale floating in the ice."

The fog was socked in solidly most of the time; it lifted only for short intervals and made matters doubly difficult for the wide-ranging boats to get back to their vessels. All this time, the ships were lying close to the shore, anchored or fast to the ice, waiting for it to open off the land as it was expected to do with the first strong northeast wind that blew.

In the meantime, there were reports of whales being seen off Seahorse Islands and several of the ships sent boats up there for expeditions of two days or longer. The crews were ordered to catch and cut up the whales on the ice, as there was no chance to get there with the ships because of the ice and shoal water.

This is probably about the most demanding kind of whaling there is. It meant that the boats' crews had to sleep either on the shore or the ice, that they had to carry extra gear in order to cut in the bowheads, and that they had to tow the blubber miles back to the ships. It was backbreaking, discouraging, dirty work under conditions of great hardship, and the crews must have wondered, every time they went onto a bowhead, whether it was better to leave him alone and go without the money or kill him and endure the wearisome labor and discomfort that would follow.

But about this time, prospects brightened. On the twenty-fifth of August, it blew a strong northeast gale and the ice opened and went offshore. During the next couple of days, the weather was good, whales were plentiful and the fleet took several. The ships, once more able to get under way and go about normal business, stood offshore and commenced whaling in earnest. The consensus was that the ice was going off for good; many of the ships were boiling, and a feeling of general optimism prevailed.

The optimism prevailed precisely until the twenty-ninth of August, at which time the light southwest winds freshened toward the end of the day, setting the ice inshore so fast that some of the ships were caught in the pack. The rest retreated inshore ahead of the ice and here they anchored in three to four fathoms of water, the main ice coming in toward them, and the small ice packing around the ships. Aboard the *Eugenia*, the mate noted dutifully: "August 29— Plenty ice. At 11 P.M., came to anchor in seven fathoms near Wainwright Inlett. Employed in boiling."

Now because men will be men, and especially because as whaling wound down in the last half of the nineteenth century it attracted lesser types from the back blocks and back alleys of half the world, there were also matters that went on in the fleet which had nothing to do with whaling. They arose from incompatibility or discontent, or both, and very likely stemmed, in part, from the season's setbacks due to weather and the frustrations and wasted effort these produced.

On the 30th, the *Eugenia*'s mate was concerned with a problem of human relations. He wrote: "At 2 finished a-boiling. Ice drifting down past the ship some heavy. At 3 o'clock, two of the men got to fiting. Mr. Coner, the 2d offerser, went forward to stop the row and one of them, George White by name, sayed that he wood not for him or eney sun of a bich. Mr. Coner took hold of him to part them when he struck Mr. Coner. He then hauled him Clear and while in the act of taking him aft, another one (James Dearman) in-

terfered with a large crow bar, saying to Mr. Coner that he wood
punch his ies out with it if he did not lett his Chum go."

How a shipboard situation such as this is handled depends on the
temperament and training of the vessel's authority. But two things
are clear at the outset, and experienced masters would agree on them:
first, this is a self-contained and domestic matter; it is not necessary
or desirable to call for outside help, even if there is any available;
and second, by whatever means, this insubordination, with overtones
of mutinous behavior, must be stopped, for discipline is at the heart
of a successful ship operation.

The fourth mate went below to fetch the first officer, who was
sleeping. The latter dressed hurriedly, came topside and observed
that the second mate was leading White aft, while Dearman, shout-
ing and jeering, was trying to incite other members of the crew to
general disorder.

The first mate went below, reporting to Captain Nye that "there
was a row," and said he thought it was best to put George White in
irons in the run (that is, below decks, aft). The master agreed, and it
was done. Dearman told the first mate that he wanted to be put in
irons, too, but the mate ignored the request and commanded him to
go forward, which he did. Sending a man forward on a ship is far
more than geographical; it is both sociological and psychological—
the after part of a ship is for the hierarchy; in navy terms, it is "officers' country," and enlisted personnel or foremast hands do not venture aft unless invited or ordered to do so. Forward is where they
belong, that is their part of the ship. Dearman was, therefore, being
put in his place, which he understood very well.

All of this obviously was before witnesses, for the rest of the crew
was watching, and it was of first consideration to preserve the order
and sanctity of the quarterdeck so that everybody would understand
completely that there was no erosion whatever of authority.

Four hours later, Captain Nye took charge of the affair. The time
element here is revealing and undoubtedly indicates that Nye was a
seasoned master of characteristic stability and common sense. In the
first instance, he concluded correctly that his mates were handling
the initial situation ably; it was his judgment that time might have

lowered the tensions somewhat, and that he could most effectively apply his supreme authority later, when it was obvious to everybody that he was acting from neither fear, weakness, nor an inability to restore order.

However, it didn't prove to be quite that simple. Nye had White brought on deck and the latter's irons were removed, at which time he embarked upon a torrent of filthy language, threatened the captain and his officers several times, and concluded by saying there would be "plenty of dark nights" before the ship returned to the Sandwich Islands, the implication being that he would throw the afterguard overboard to drown, given the opportunity.

In response, Nye ordered White triced up in the mizzen rigging, an ingenious deterrent in a fleet in which flogging and similar punishments were not generally employed.

Then Nye sent the first mate for Dearman, who was brought before him. According to the log, Dearman was "saucy and incelent, also making maney threats and thretning Captain Nye and the offercers that he wood doo this and doo that." So he was tied up in the rigging, too.

According to the mate, "He (Dearman) then told his frend White to hold out but he soon began to think different of it. When they was up there twenty minutes they began to want to be lett down analazing that they was in the rong and baging Captain Nye and the offercers pardon."

However, the mate let them hang there for a half hour before he finally cut them down. He ordered them to walk the deck awhile to cool off.* While they walked, both men immediately began to make further threats and were strung up for another twenty minutes after which they "promest to be good men and make no more trouble and doo thear duty." The mate was not at all sure they really would, but he was comforted by the fact that their discontent was

---

* I once interviewed an old boatsteerer who had sailed on New Bedford whaleships and when he said the captain of his ship had triced up a dissident crew member, I asked him precisely how it was done. He replied, "Our Old Man had him tied up by the thumbs just so his toes touched the deck when the ship wasn't rolling and every time she rolled, his toes wouldn't touch, and he would swing by his thumbs and he quick got damned tired of that."

not widely shared in the forecastle. He wrote, "Thar 2 vere Bad men as we have good reasons to now. Thay have been talking about thare knives Sevrell times before if thare wase truble also that thay had the ship whare they wanted her, that the offersers were afraid of them and that they wood doo as thay pleased.

"This Dearman when he was colled aft he called on the men forward to come and help him whitch they wood not doo as thay have been a-rowing with thare shipmates they have gut no more freinds forward than thay have aft."

Such problems were neither common nor uncommon; in the present instance, they were undoubtedly aggravated by the unusual perversities of the season. In a five-day period, for example, the *Emily Morgan* lowered and struck a whale with two boats, losing him when one of the lines parted and an iron drew; sent five boats into the ice, struck another whale, losing him and the line as well; saw a number of whales, struck one, and lost him; and sent boats back into the ice in a heavy fall of snow, hunted for hours, and saw nothing, the crews returning to the ship weary and half-frozen.

Meanwhile, the ice continued to drive in around the ship, forcing Captain Dexter to maneuver continually, hauling the ship out to the ground ice, making her fast, anchoring in the narrowing clear water left, getting up the anchor and working offshore the little that the ice would permit, over and over.

"While catting the anchor," wrote the *Morgan*'s mate, "Frazier, a boatsteerer, having too much to say, was told by the first officer to keep silence, which he would not obey. The order being repeated several times without a compliance, the mate was compelled to use force to preserve the discipline of the ship and in doing so, accidentally broke Frazier's jaw. This is not the first time Frazier has refused to keep silence when ordered. Otherwise, he is an excellent man."

At this point, late in August, the fleet had been driven by the ice into a narrow strip of water not over half a mile in width at its widest part. Here, scattered along the coast for twenty miles, they lay, the water from fourteen to twenty-four feet deep and ice as far as the lookouts at the masthead could see.

The bark *Monticello*, Captain Thomas W. Williams, having been ashore three days when set down into shoal water by wind and ice working toward the land, came off into deeper water without damage.

# 15.

# "Our dreadful situation"

Up to this time [August 25], no immediate
danger had been anticipated by the captains beyond that incidental to their usual sojourn in
these seas. The Esquimaux, nevertheless, with
the utmost friendliness, advised them to get
away with all possible speed as the sea would
not open again. But as this was contrary to the
Arctic experience of the whalemen, they resolved to hold their positions.
—*Harper's Weekly*, December, 1871

The fact was that the northeast gale of August 25 drove the ice
offshore no more than eight miles at best and did not offer the whaling fleet opportunity to "get away with all possible speed," even had
its masters chosen to do so.

This warning was late in the day, but it was not the first warning
by the natives. Early in the season of 1871, when the Eskimos came
out to the ships, they said that the weather would be unusually bad
and that if the whalers ventured in as usual between the ice and the
land, they would not be able to get their vessels out.

The prevailing wind in summer on the northwest coast of Alaska
is from the east and this works the ice off the land and disperses it.
Each year, the fleet was accustomed to working up by the land in
the water thus cleared, ordinarily finding plenty of whales there.
By the close of the season, when the northwesterly winds became
prevalent, the ice usually had become so broken up and melted that
it had ceased to become an element of danger to shipping. At that
time, the vessels were compelled to retire to the south by heavy ice
drifting along the coast, rather than by a threatening closing in of the
pack upon the land.

But even by the end of August, the season of 1871 had already been markedly different in pattern. The easterly winds were not so strong and constant as usual, and the ice that had gone offshore returned in a pack so heavy that it was impossible to get a ship through it or even to hold against it with the anchor down. What we are concerned with here are heavy fields of freshwater berg ice, not icebergs of the immense proportions seen in the Greenlands seas, but masses so solid, nevertheless, as to ground in ten fathoms of water.

The whaleman did not understand much of the Eskimo's language, and regarded him as a simple, primitive, superstitious child. More to the point, a whaling master could not very well explain to a Howland, or anybody else at home in the New Bedford countinghouse, that he had decided not to go bowheading because Toonook, the bad spirit, was especially displeased about something. If New Bedford's whaling merchants had declined to take seriously the whaling master's thought in 1852 that perhaps God did not want his bowhead chased into the ice, they were not at all likely to take Toonook seriously some twenty years later.

But the Eskimo's warning, so casually dismissed—even if it was based in part on nonscientific factors—had scientific support. The winter of 1870 had been, according to Charles Bryant, special agent of the U.S. Treasury Department, "most remarkable for the long continuance of the cold and its unusual severity . . . as shown by the records kept at the Seal Islands, Port St. Michaels, Norton Sound, and the posts on the Yukon River. The ice floe came down on St. Paul's Island February 5 and continued until May 5, ninety days, the usual time of its duration being from five to twenty days in April. It may have been the Eskimos' knowledge of this that induced them to warn the vessels of their danger. Under ordinary circumstances, this is not likely to occur again."

Beginning on August 29, the circumstances of the Arctic fleet of 1871 began to deteriorate rapidly. The trap began to close.

When the northwest wind of this place blows seriously, it is in the form of a roaring, undisciplined gale off the roof of the world, a fierce and bitter blast driving needles of thick snow, and there is no

standing against it. It is an irresistible force kin to the globe's beginnings; it whistles through the Arctic's half-empty world and threatens to empty it. It drives life out of sight and once unleashed by the
deteriorating season persists unceasing for almost more days than
the human spirit can endure.

And where there is open water, the sea is tortured by the wind.
Distressed by the gale, the waves assume an attitude of formidable
heaviness; as they approach, dark and rising, only the ugly undula-
tion of their crests betray that they are liquid, not solid. Yet they
are close to solid, for all that, and it is impossible to observe them
rolling down the wind, staggering and building, without being conscious of the danger—perhaps even hostility is not too poetic a
word—that they represent.

To be to leeward of such seas is to be immediately conscious of
the fragility of man's strongest works. When these waves hit anything short of solid rock, it shudders, is overwhelmed or smashed,
or both. The water, its hammerblow delivered, sprays into instant
ice, the cold flow arrested in its torrent and downcoming; it encrusts
bow, stay and spar, immobilizes ship and man, a terrible, crystal
monument to indiscriminate power indiscriminately applied.

And where ice covers the sea, it is in chunks bigger than houses,
ponderous blocks, gale-driven, that come charging toward the land,
and nothing will halt their assault until they ground in the shoal
water. And when they ground, with wind and sea still pounding
them, they pile up, smash and grind like millstones; they pack until
they build massive ridges of points, peaks, and angles, of shattered
planes, drunkenly stacked, all shoved against the low and turbulent
sky, testimony to the limitless force of the wild atmosphere that was
their architect.

On the twenty-ninth of August, 1871, a fresh breeze from the
southwest drove the ice back inshore with such rapidity that several
ships of the fleet were caught in the pack. The remainder of the
vessels—some had to slip their cables in order to move quickly
enough—retreated ahead of the incoming ice and anchored in three
to four fathoms of water, the ice still coming in and packing around

them. The heavy floe-ice grounded in shoal water, and between it and the shore lay the ships, with scarcely room to swing at their anchors.

Even from the masthead, there was nothing but ice visible offshore, a jumble of closely packed chunks that stretched, white and cold, right up to the skyline. The only clear water was where they lay and that was narrowed to a strip from two hundred yards to a half mile in width and extended from Point Belcher to two or three miles south of Wainwright Inlet. The southeast and southwest winds continued, light from the former and fresh from the latter direction, and with each day's passage, the "small ice" packed more and more closely around the beleaguered ships.

On the first day of September, the thick ice came down upon the *Eugenia* with such force as to part her anchor chain and drive her toward the beach. The crew succeeded in getting her other anchor to hold, three-quarters of a mile from land, in seven fathoms. Her mate, noting that the incident had cost them an anchor and 180 feet of chain, observed, "Some twenty-six sail in sight. All jammed in the ice close onto the beach. Things look bad at present."

Abruptly, they looked worse. At three o'clock the next morning, the watch aboard the *Henry Taber* observed that the Hawaiian brig *Comet*, lying nearby, had set her ensign at half-mast. Captain Packard of the *Taber* ordered a boat lowered immediately and set off for the *Comet* to find out what was wrong.

The closer they got to her, the more obvious her plight became. The brig was squeezed between two massive chunks of ice that had forced her upward, half out of water. She was being pinched so hard that her stern was forced out, shattered, and driven from its heavy fastenings just as pressure forces a cork out of a bottleneck. Her crew was in the ice, in their boats, and her master, Captain Sylvia, told Taber and others who came in response to his signal that every timber in the ship had been snapped by the crushing.

It was snowing hard, thick flakes slanting across the dark water so heavily as to reduce visibility to virtually nothing. Because the ice that held the *Comet* was motionless, either packed or grounded, the boats' crews took a chance, boarded her and salvaged what they could. Packard got a cask of bread, one of flour, a whaleboat, and invited Sylvia to take passage with him. The *Comet*'s crew was split up among the fleet. The brig hung, suspended, smashed, and helpless, for more than three days—a specter for them all—and when tide and current moved the ice that held her in its grip, she sank.

Still, there was some business as usual. Off the mouth of Wainwright Inlet, the *Thomas Dickason*, one of George and Matthew Howland's vessels, edged into a favorable hole of open water and cut in a whale that was in the ice, probably one ironed and lost a day or so before by the *Eugenia*, which could not get to it. Aboard the *Eugenia*, the mate wrote in the log that they "cood not gett to him to see whose whale it was but rather suspisios that it's ours."

If it seems strange, in retrospect, that the fleet continued to go whaling, even in a deteriorating situation, it was paradoxically due to the experience of the masters, not lack of it. They were still expecting a northeasterly gale that would drive the ice offshore and allow them to work easterly along the land, as they had in every season before. There was not one among them who could remember a time when an eventually favoring wind had not allowed them to pursue the whales and get out again before the winter weather shut down. Even at this point, the whaleman was more uneasy over loss of precious time in an inevitably short season than he was over the weather.

The bark *Roman* was next. For some days, she had drifted, in due course arriving off the Sea Horse Islands, where she took a whale. It was brought alongside, and she was cutting in, when she was caught between two floes that came down upon her.

The crew went over the side onto the ice in a matter of minutes, realizing that their position was desperate and that they had no chance to take gear or provisions with them. They watched, stunned and shaken, shivering on the open floe in the blast of weather as the ice caught the ship on each side, lifted her bodily, keel out, and then began a terrible process of squeezing and relaxing. Three times the jaws came together upon the *Roman*. Then she disintegrated and went to the bottom. Forty-five minutes from the time the ice first closed upon her, there was nothing to show where she had been.

The *Roman*'s company went to other vessels. Captain Edmund Kelley of the *Seneca* came down in a whaleboat from the northern part of the fleet to report that more than one-half of them were fast in the ice and the others crowded nearly to the beach. The *Dickason* took two more whales in the same hole about the fourth of September and the *Eugenia*'s mate went to the masthead to look at the northern fleet. "Some twenty of them," he said. "One of them a-boiling."

A couple of days later, a fresh southeasterly set the inshore ice in motion, but not in any direction to help them. The *Eugenia* was squeezed, losing considerable of her bottom copper and some sheathing. At four in the morning, the *Taber* lowered all her boats and went whaling; they probably would have struck one except for the fact that the chief mate of the *Florida*, angered because he was not going to get the bowhead himself, drove the whale down by whacking the water with his steering oar. The *Taber* gave up temporarily, drifted to the southwest about two miles, and anchored with several other ships. Her mate wrote in the log, "It looks like a poor show to get out of this this season."

By running out her singled-fluked, light ice anchor and "using up all of the rope in the ship," the *John Wells* was hauled out of the ice in an effort that took her crew the whole forenoon. This operation involved finding an ice chunk that was grounded and big enough to be stationary, chopping a hole in it with an axe, hooking the anchor into it, and running a line from the anchor to the ship, where it was wound up on the windlass—in this case, tediously, by persistent manpower—forcing the ship ahead. The men of the *John Wells* were rewarded for their labors by having her go ashore, where she still lay at eleven that night when they gave up and went to bed. They finally got her off the next morning and promptly set about trying to help the *J. D. Thompson* which lay nearby, hard and fast on the bottom.

At 4 A.M. on the seventh of September, the *Emily Morgan* saw whales and sent her boats off. The bow boat struck, the whale ran for the ice, and before he could get under it, the waist boat succeeded in fastening its line to the one he was towing so they would

SEA HORSE ISLANDS.
ROMAN.                           POINT BELCHER.

GAY HEAD. GEORGE.

PLATE I of Benjamin Russell's series of five lithographs titled "Abandonment of the Whalers in the Arctic Ocean, Sept. 1871." *Courtesy of The Whaling Museum, New Bedford, Mass.*

not lose him; the whale appeared hard-hit and they thought he would die soon, which proved to be the case. But the taking was marred with tragedy; the second officer, Antonio Oliver, while handling his bomb gun, accidentally shot himself. The bomb entered midway of his lower jaw near the throat and came out at the back of his head, killing him almost instantly. One boat took the dead man back aboard the ship; the others towed the whale into clear water. The *Morgan* came alongside and anchored about one thirty in the afternoon and began to cut in.

The *Taber*'s boats found two or three whales that day but they were very shy and the boats never got close to them. There was a strong southeaster with snow and rain during much of the day and the large ice did not move at all. "Some twenty-three ships in sight," the *Taber*'s mate wrote, "and nearly every one in the fleet despairs of getting their ships south this fall, which is a cause of much uneasiness with everyone."

Also on the seventh, Captain Owen of the *Contest*, with Captains James H. Fisher and Dexter, of the *Oliver Crocker* and *Emily Morgan*, went in their boats to sound along the ground ice to the southward, to determine the depth of water and evaluate the chances of escaping with the ships. The results of their survey were as depressing as the miserable weather in which they worked. They found less than nine feet of water in several places south to the shoal off Wainwright Inlet and no more than four to eight feet across the shoal, off which were now lying the *Dickason, Awashonks, Julia,* and *Eugenia.* There was only one conclusion. It was impossible to move most of the ships to the southward toward freedom because they drew far more water than was available in the only channel still open.

By the eighth of the month, the pretense of whaling was ended and no one denied any longer the peril that threatened. On that day, the bark *Awashonks* was destroyed by the crush of ice, in somewhat the same manner as the *Roman*, except that she was driven into shallow water and did not sink. Instead, she lay there, wrecked and dismasted, for them all to see and ponder. A third shipless crew was distributed among the remaining vessels of the fleet, and Captain

Daniel B. Nye of the *Eugenia* bought what was left and salvageable of the *Awashonks* for sixteen dollars.

There was absolutely no improvement in the weather. Every day brought thick fog that blotted out both barren land and frozen sea and in the motionless and dimensionless microcosm in which the whalemen were trapped, it was increasingly evident that the end of the season was approaching. The ice gave no sign of starting off-shore; on the contrary, the land water did not now exceed a half mile in width and no more than half that distance was navigable. Such little clear water as remained was rapidly filling with small ice that packed in around them.

September 8 was a day of both difficulty and decision. The *Seneca* recorded "thirty ships in sight and all crowded into shoal water by the ice, with every prospect of being drove ashore in standing in-shore to clear the ice." As early as four in the morning, a gale roared in from the south and belted the fleet as it lay inside the grounded ice barrier in three fathoms of water or less. It not only destroyed the *Awashonks* but drove ice in upon the *Eugenia*, parting her rudder tackles, slamming her rudder hard against the stern and smashing its stock, and carrying away the hardware that held rudder to sternpost —all of which meant that she could no longer be steered. The bark *Elizabeth Swift* was grounded, and it took hours to get her off again. At two in the afternoon, a boat's crew from the *Morgan* took the body of Mr. Oliver, the second officer, ashore. The mate read from the Bible over the half-frozen hole they had dug, and they buried him.

Even as this service was going on, the masters of the whaleships held the first of a series of conferences aimed at estimating their prospects and evaluating the options that were open to them. The burden upon them was both immense and unprecedented. Captain Kelley of the *Gay Head* wrote, "We felt keenly our responsibility, with three million dollars worth of property and 1,200 lives at stake. Young ice formed nearly every night and the land was covered with snow. There was every indication that winter had set in."

It was immediately obvious that if all or most of the ships could

JOHN WELLS. MASSACHUSETTS.    J. D. THOMPSON.    CONT

HAMPION    E. SWIFT  O. CROCKER   REINDEER  G. HOWLAND  CARLOTTA.
RGAN.  HENRY TABER.    FLORIDA.    NAVY.   SENECA.    FANNY   PAIEA.

PLATE II. "Abandonment of the Whalers in the Arctic Ocean, Sept. 1871." *Courtesy of The Whaling Museum, New Bedford, Mass.*

not be gotten to clear water that was an unknown number of miles away—and the soundings already taken indicated that they could not —the more than one thousand people aboard them somehow had to be removed to safety in the very short time remaining before even that became impossible. The fleet, prepared only for the normally short whaling season, had provisions for no more than ninety days. Milder weather was at least nine months away, and to try to survive the winter ashore was to invite wholesale death from starvation and exposure. Every shipmaster, every mate, every man in the forecastle knew in detail what the winter before had been like for the survivors of the *Japan*. All were well aware that the land offered no sanctuary, the generosity of the impoverished Eskimo notwithstanding, and that the hostility of the environment would kill most of them if they did not get out.

Over the next couple of days, they agreed upon two courses of action, and shipmasters were assigned to pursue them immediately. The first concerned the Hawaiian whalers *Kohola* and *Victoria;* they were the smallest ships in the fleet and, therefore, drew the least water. There was some hope that either one or both could be lightened sufficiently so that they could make their way south in the shallow land-water channel. If this worked, they could transport some of the ships' crews to open water and try to make contact with an unknown number of whaleships that it was hoped were still within reach to the south.

The second action involved sending several masters in whaleboats south along the land to discover how far it was to open water and to see whether there were, in fact, any other ships in the area that were not caught in the ice. The bleakest possibility was that any ships south of them, discouraged by winter's unseasonable onset, might already have started for home. If this proved to be the case, there was no help for them, no help at all, and they were staring into the face of death.

Under the circumstances of depression, rotten weather, a consistently turbulent sea that impaired shiphandling, and the necessity of haste, it was backbreaking labor to hoist the casks of oil and provisions out of *Kohola* and transfer them to the deck of the

*Carlotta* of San Francisco that had been maneuvered alongside. Periodically, anxiously and yet with rapidly diminishing hope, they checked *Kohola*'s draft and after hours of effort, had to concede in discouragement that she just was not coming up enough. When they had virtually everything out of her that they could take out, she still drew nine feet and there was no more than five and a half feet of water over the Wainwright Inlet bar.

Captain Kelley of the *Gay Head* and Captain Williams of the *Monticello* had no better luck in trying to get the *Victoria* over the shoal. It was clear that they had no hope of getting the ships out, not even one ship, and if there was salvation for them, it had to lie in some other direction.

On the ninth, the mate of the *Morgan* wrote, "Offshore is one vast expanse of ice. Not a speck of water to be seen in that direction. All but three of the northern fleet have come down and anchored near us. There are twenty ships of us lying close together. There seems to be but little hope of our saving the ship or of any of the other ships being saved. Commenced making bags for bread in case we have to abandon ship."

The weather remained miserable—cold, thick, and windy, with increasingly frequent snow squalls. In one overnight period, the young ice around the ships made from one to three inches. Every day, the vessels were packed in more solidly. The ship *Julian* was "leeking quite bad." Captain Packard of the *Taber* left early on the morning of the ninth in his whaleboat to determine how far south it was to clear water; fifteen hours later, he got back to the ship and reported that he had had trouble breaking way for his boat through the ice of the land water and that he had made ten miles before turning back, without seeing any open water offshore.

Captain Kelley and Captain Williams also went south in their boats, and as Kelley subsequently wrote, "The search for open water was in vain. We felt there was no possibility of rescue . . . we were anxious and meetings of all the captains were held nearly every day and every phase of the situation was thoroughly discussed."

The prospect of hopeless isolation grew larger. Suppose there was no open water within any reasonable distance? Suppose there were

no ships free of the ice? It seemed inevitable that they would have
to abandon their ships which each day faced the narrowing choice between being crushed in the ice or wrecked on the shore. Even assuming there was a sufficient channel for the whaleboats to reach
the sea beyond the ice, if there were no ships there, what then?

"For my part," Captain Dexter of the *Morgan* declared, "I will
not cross the Arctic Ocean in an open whaleboat laden with men
and provisions in the latter part of September and October. As far
as Icy Cape, there is no danger, but beyond that, if all the ships'
companies have to take to boats to Bering Strait, the sea is dangerous
at this season of the year. Out of the twelve hundred, not a hundred
will survive." It was Dexter's view, and many of his colleagues
agreed, that if rescue ships could not be found, as bad as the prospects were, they would return to their icebound commands. The
prospect of death ashore was slightly better than that of death crossing the Arctic Ocean in winter.

At this point, an expedition of three whaleboats, under the command of Captain D. R. Fraser, of the *Florida,* was sent down the
coast to determine how far the ice extended, what chances there
were of getting through the barrier, what vessels, if any, were outside the ice, and what relief from such vessels might be counted
upon.

The first question, of course, was whether the whaleboat expedition could fulfill any of its assignments. If the ice to the south of
them had already come in upon the land, no channel, even for a
shallow-draft boat, would be available to them. Even in the area
around the ships there was now considerable difficulty in moving
about with a whaleboat. On the forenoon of the tenth (the day
Fraser left), the *Reindeer* and *Contest* weighed anchor and for
nearly an hour scarcely moved, so firmly had the ice made around
them.

Aboard the *Morgan,* the mate wrote: "The wind, though favorable, has not moved the ice. The land water is frozen completely
over to the thickness of an inch so that it is difficult for whaleboats
to pull through it and it retarded progress of the ship a great deal
when we weighed anchor and took a berth a mile farther to the

southwest. There is no hope of saving the ship and we are preparing to abandon ship. The only hope of saving our lives is that the ice in the land water will break up." But even if it did, the simple fact that they would be able to go down the coast in whaleboats did not mean that their lives would be saved.

On the tenth, the *Kohola* was aground. At the meeting of masters, it was decided to freight some provisions south—it was hoped as far south as Icy Cape—if the whaleboats could still get through, storing the food ashore so that they would have something to live on even though the ships were destroyed or abandoned. Throughout the fleet, they were breaking out bread and other foodstuffs, boiling pork and beef in the trypots and stowing it in barrels for the trip down the coast.

All hands were fitting out their boats for the trip. To protect the bottoms against ice, they sheathed them with copper, for ice will cut wooden boat planking like a knife. To strengthen them for hauling out on the rocks and ice of the shore, they fashioned false keels. And to keep out the water—knowing the boats would be heavily loaded—they built risers on the gunwales, to increase the height of their sides.

The weather remained foul. As the hours passed, there was rising concern because they had heard nothing from the Fraser expedition and the first boats sent down the coast to land provisions had not returned, although their crews were well aware of the need for haste. It was possible to take two views of this silence—either the boats could not get south or, having made it, could not get back to the fleet because the incoming ice had cut them off.

But the eleventh brought the first good news, even though it did not come from Captain Fraser. Aboard the *Emily Morgan,* first officer William Earle wrote, "Sets in rainy. Broke out bread and provisions. Boiled six barrels of beef for the boats. Two boats ready for a start. The ice is slowly narrowing the already narrow strip of water and rendering our position extremely dangerous.

"All of the ships are preparing to send provisions south to subsist on till some ship can be communicated with. Furnished two boats with provisions and clothing for the crew for a start southward."

Here the mate was interrupted. Then he resumed. "As I write this, one of the boats that started two days ago has returned, reporting that there are three ships in clear water and a prospect that four more will soon be in safety."

The report spread rapidly through the fleet, although not necessarily accurately. Aboard the *John Wells*, Nathaniel C. Ransom of Mattapoisett wrote in his journal, "Got news from ships to south'rd. They seem to think they are safe enough as yet." Actually, at the time at which he made this entry, no such optimism was justified. The ships to the south were having their problems. They were not "safe enough as yet." As a matter of fact, several were in trouble. And, even if they had not been, they were, in terms of aiding the icebound whalemen off Wainwright Inlet, many hard miles away.

On the twelfth, the vigorous southwesterly wind persisted, continuing to drive the ice toward the shore. The log of the *Seneca* noted, "We have now just room to swing to our anchor clear of the land and the ice." At four in the morning, Captain Dexter of the *Morgan* left in the starboard boat to take his wife "to a place of safety to the south," but said he would return from Icy Cape if no ships were to be found. A stream of whaleboats, including those from the *Morgan, Contest, Seneca, Concordia* and *Gay Head*, worked down the land water carrying bread, meat, and flour.

Captain Fraser returned and a meeting of the masters was summoned immediately. He reported what everyone had come to assume, that it was utterly impractical to try to get any of the main body of the fleet out. He had found that the whaler *Arctic* and another vessel were in clear water below the icefield, which extended to the south of Blossom Shoals, approximately eighty miles away, and that five more whaleships, then fast in the lower edge of the ice, were likely to get out soon.

The masters of the doomed whaling fleet then drafted in conference two documents, poignant in their austerity, historic in their importance, and remarkable for their brevity, considering how much had to be said.

It is no more than fair to them to record the moment more fully than they chose to do. It is only proper to reemphasize why it is, these generations later, that we have no more than a barebones record of what must have been for them—even though they were accustomed to a career of hard choices—an excruciating, deeply saddening interlude for which they knew they would have to answer to the world, and especially to their peers, if they lived.

Logbooks are not supposed to reflect emotions; that is not what they are for. Most shipmasters did not keep journals and even if they did, were not by nature inclined to bare the soul. Their careers were bedded in self-discipline. This control, just as lack of it would have been, was contagious. It is easy enough to say now that panic would not have done anybody in the Arctic fleet of 1871 any good —it never does anybody any good. But the impressive fact is that nobody, including the handful of women and children, is recorded as having lost his composure. Even such work as could be done was done as well as possible, as if they were not going to abandon ship at all.

Yet among those masters, seasoned men all, facing each other grimly in the after cabin of the *Champion*, there must have been unrevealed thoughts and emotions, including heartbreak, as fierce and wild as the weather itself that forced them to do what they did.

This is what they had to say:

*Point Belcher, Arctic Ocean, September 12, 1871*

Know all men by these presents that we, the undersigned, masters of whaleships now lying at Point Belcher, after holding a meeting concerning our dreadful situation, have all come to the conclusion that our ships cannot be got out this year and there being no harbor that we can get our vessels into and not having provisions enough to feed our crews to exceed three months and being in a barren country where there is neither food nor fuel to be obtained, we feel ourselves under the painful necessity of abandoning our vessels and trying to work our way south with our boats and if possible get on board of ships that are south of the ice.

We think it would not be prudent to leave a single soul to look after our vessels as the first westerly gale will crowd the ice ashore and either crush the ships or drive them high upon the beach. Three of the fleet have already been crushed and two are now lying hove out, which have been crushed by the ice and are leaking badly. We have now five wrecked crews distributed among us. We have barely room to swing at anchor between the pack of ice and the beach and we are lying in three fathoms of water.

Should we be cast on the beach, it would be at least eleven months before we could look for assistance and, in all probability, nine out of ten would die of starvation or scurvy before the opening of spring.

Therefore, we have arrived at these conclusions after the return of our expedition under command of Captain D. R. Fraser of the *Florida*, he, having with whaleboats worked to the southward as far as Blossom Shoals, and found that the ice pressed ashore the entire distance from our position to the shoals, leaving in several places only sufficient water for our boats to pass through and this liable at any moment to be frozen over during the twenty-four hours, which would cut off our retreat, even by the boats, as Captain Fraser had to work through a considerable quantity of young ice during his expedition, which cut up his boats badly.

[Signed by the masters]

The following is an incomplete list of the masters who signed this document—incomplete because neither the original document nor a full copy has been located and because some of the vessels were out of San Francisco or Hawaii and New Bedford records do not include them:

Robert Jones, *Concordia;* Leander C. Owen, *Contest;* Benjamin Dexter, *Emily Morgan;* Timothy C. Packard, *Henry Taber;* Hezekiah Allen, *Minerva;* James H. Fisher, *Oliver Crocker;* Lewis W. Williams, *Fanny;* Abraham Osborn, *George;* Aaron Dean, *John Wells;* George F. Bouldry, *Navy;* Valentine Lewis, *Thomas Dickason;* Ariel Norton, *Awashonks;* William H. Kelley, *Gay Head;* West Mitchell, *Massachusetts;* H. S. Hayes, *Oriole;* B. F. Loveland, *Reindeer;* Edmund Kelley, *Seneca;* George W. Bliven, *Elizabeth Swift;* Thomas W. Williams, *Monticello;* ——— Redfield, *Victoria;* D. R. Fraser, *Florida;* James H. Knowles, *George Howland;* George F. Marvin, *Daniel Webster;* Henry Pease, *Champion;* Daniel B. Nye, *Eugenia;* and ——— Sylvia, *Comet.*

The "Captain Sylvia" of the *Comet*, mentioned in the *Henry Taber*'s log, may have been Joseph D. Silva, listed in New Bedford records as a whaleship master in this period. Jared Jernegan may

have been master of the *Roman* and a Captain Allen, first name un-
listed, was the captain of the *J. D. Thompson* in 1868. (Several of
the above masters listed had assumed these commands in that year.)

The second document prepared by the masters was an appeal for
help. In its implicit assumption that the reply to the appeal will be
affirmative, it provides a poignant clue to the almost universal willingness of one sailor to help another.

—Ship *Champion*
*Off Point Belcher, September 12, 1871*

To the masters of the ships in clear water south of Icy Cape:

Gentlemen: By a boat expedition which went out to explore the
feasibility of a ship's passage to clear water, report there are seven vessels
south of Icy Cape in clear water sailing.

By a meeting of all the masters of the vessels which are embargoed
by the ice along this shore, as also those that have been wrecked, I am
requested to make known to you our deplorable situation and ask your
assistance. We have for the last fifteen days been satisfied that there is
not the slightest possibility of saving any of our ships or their property,
in view of the fact that the northern barrier of ice has set permanently
on this shore, shutting in all the fleet north of Icy Cape, leaving only a
narrow belt of water from one-quarter to one-half mile in width, extending from Point Belcher to south of Icy Cape.

In Sounding out the channel, we find Wainwright Inlet to about
five miles east north east from Icy Cape the water in no place of sufficient depth to float our lightest draught vessel with a clean hold, in
many places, not more than three feet.

Before knowing your vessels were in sight of Icy Cape, we lightered
the brig *Kohola* to her least draught, also brig *Victoria*, hoping we
should be able to get one of them into clear water to search for some
other vessel to come to our aid in saving some of our crews. Both vessels
now lie stranded off Wainwright Inlet. That was our last hope, until
your vessels were discovered by one of our boat expeditions.

Counting the crews of the four wrecked ships, we number some
twelve hundred souls, with not more than three months provisions and
fuel; no clothing suitable for winter wear. An attempt to pass the
winter here would be suicidal. Not more than two hundred out of the
twelve would survive to tell the sufferings of the others.

Looking our deplorable situation squarely in the face, we feel convinced that to save the lives of our crews, a speedy abandonment of

our ships is necessary. A change of wind to the north for twenty-four hours would cause the young ice to make so stout as to effectually close up the narrow passage and cut off our retreat by boats.

We realize your peculiar situation as to duty, and the bright prospects you have for a good catch in oil and bone before the season expires; and now call on you, in the voice of humanity, to abandon your whaling, sacrifice your personal interest as well as that of your owners and put yourselves in condition to receive on board ourselves and crews for transit to some civilized port, feeling assured that our government, so jealous of its philanthropy, will make ample compensation for all your losses.

We shall commence sending the sick and some provisions tomorrow. With a small boat and nearly seventy miles for the men to pull, we shall not be able to send much provisions. We are, respectfully yours, Henry Pease Jr., with thirty-one other masters.

Miles to the south, Nathaniel Ransom of the *John Wells* wrote in his journal, somewhere on a frozen beach: "This morning, we started south with a load of provisions. Evening, tied up ashore for the night."

And the mate of the *Taber* wrote in his log: "Our beautiful *Henry Taber* lying at anchor close to the beach and close to the ice. This morning, sent off three boats with two casks of bread and one of flour and seven barrels of meat. Journeying south to get south of the ice. Finished fitting our boats and making preparations to flee for our lives."

# 16.

# Abandonment

September 13—Several meetings were held today to discuss affairs and it was decided to fix
tomorrow, at 10 A.M., to abandon the ships.
—CAPTAIN LEANDER C. OWEN,
of the ship *Contest*

The morning of the thirteenth was an anxious time. It was essential to break out and prepare more provisions, outfit and improvise
more boats for the trip down the coast, in order to insure the best
chance of survival. Yet the question hanging over their heads was
whether they had already waited too long.

About a hundred whaleboats were somewhere between the iced-in
fleet and Icy Cape. Although they had been gone a day or two, only
a few had been heard from. They had been dispatched to transport
food down the coast, together with some few personal belongings,
and to put these provisions aboard any ships found in open water, if
they could get to them. "As yet no report from the boats," wrote
the mate of the *Seneca* gloomily, and the log of the *Morgan* concluded that, since the boats had intended to return for men and more
provisions, "this is an indication that the ice has closed in far to the
south."

The weather remained as gloomy as the prospects. And it was
getting colder. The night of the thirteenth was miserable for all of
them, of whatever station.

For the masters, the die was cast. It is not too much to suggest
that when a man whose principal reputation rests upon his ability
to be responsible for a ship is forced to abandon the ship—especially
if it is intact and undamaged as many of these were—it leaves a mark.
Somewhere in the back of the mind, there had to be persistent doubt.

For what was about to happen—the abandonment of a fleet—defied tradition and established precedent. Had everything been done that could have been done? Had their judgment been sound, and would their peers support their decisions? Would one more day, or even two, make a miraculous difference, bring the brisk northeasterly that would drive the ice off the land as they had seen it occur year after year?

From the deck, because the immovability of the elements in the equation of catastrophe—land, ice, and water—was so obvious, the rightness of the decision was equally obvious. Belowdeck, especially for the master, and especially on that last night, it must have been much harder to accept. In this after cabin, still snug, tight, and orderly, there is this man's home, office, world, everything. The brightly polished lamp swings easily in the gimbals, its yellow light warm on the clean white paint. The fore-and-aft table, stout mizzenmast rising through it, is, as always for a thousand days on end, solid as the Gospel, and scrubbed smooth as glass. On the forward bulkhead, the brass clock—hung there the day she was launched—ticks sweetly and records the hour as if there were no end of time.

How many hours of sleep have there been within that bunk? And the pipe, the books, the ebony parallel rules that had been a Christmas present on one of the few Christmases at home. What of them, and the hours of life they represented? How could it be that this formidable structure of wood, metal, memories, experiences, this stout citadel of knees, timbers, planks and bolts buttressed by a lifetime of sound judgment, should now prove so frail?

This is not a night for sleeping. The time is past for judgments. There is, as always, nothing but liability in maudlin thoughts, and too few hours remaining for it anyway. The work of getting ready to leave goes on.

For the foremast hands, it was no better a matter; different, but no better. Undoubtedly, they had their differences as to whose fault this was. Traditionally, in the fleets of the world of all ages, foremast hands always know whom to blame—but most importantly, they knew it wasn't theirs. There is some consolation in that because, even without knowing the statistics involved, except in a general way, they knew how much money was being left behind and lost.

This had a very personal application. Instead of a short, busy season that would have put money in their pockets, they had put in their time and effort for nothing and might better have gone whaling somewhere else or, in the case of the Kanakas, have stayed home farming. They also knew that much of what they had aboard— gear, souvenirs, and clothing—would have to be abandoned, that there was no more than room in the boats for food and people.

But beyond all this, as steady-nerved as most of them were and had to be in such a business, it was necessary to wonder whether abandoning the ships was going to make matters better or worse. Suppose the ships to the south decided not to take them aboard, but to continue whaling?

The question was something the crews of the trapped fleet could not answer. It is well enough to talk about the tradition of the sea and sailors standing by sailors, but by the last quarter of the nineteenth century, whaling had gone downhill enough, and picked up enough nonsailors, who never had heard of the sea's traditions and cared less about them, so that one could not tell what a vote would bring. And even if the masters made the decision by themselves, without benefit of democratic process, there was no telling which way the matter would go. Any foremast hand knew that some masters were miserable pinchpennies; some were so uncertain of retaining their commands, because of age, ill health, or poor luck that they thought of making money for the owner-agent ahead of everything else, and some would simply not have thought well of cramming into their ships three or four times the number of people the vessels were designed to accommodate and thus implicitly assuming responsibility for their lives or deaths.

The continuing bad weather was an adverse factor in itself. Even if the ships were still there, and still free, getting aboard them could pose severe problems, since you were talking about more than a thousand people. Oh, some of them would make it if the weather wasn't bad enough so that the ships couldn't hold on in the face of it, but if you have ever tried to bring a boat, heavily laden at that, alongside a ship with a gale blowing, and a heavy sea—and they all had—you know that some of them, perhaps a lot of them, just wouldn't make it. There are so many things, quick, simple and easy things, that convert an anxious moment into instant death and disaster. If, through a snapped oar or somebody missing a stroke in the slop and spray, the boat swings broadside in the trough, she may well be swamped in the next crest. If the sea is running so hard and heavy that you can't do a decent job of fending off when lying alongside the ship to get your people off, one good crack and you lose the whole side of the boat in a mess of splinters and everybody is overboard. Or, just as easily, if the boat rises on a crest too rapidly and slams against the ship, whoever is scrambling up the side gets smashed beyond repair. And anybody overboard in ice water, even if well-clothed and vigorous, more often than not doesn't last long enough to be picked up.

Rescue, even if there were somebody to the south willing to try it, was no sure thing. Everybody knew it.

The morning of the fourteenth brought the day of decision, but it also brought some slight reassurance because some of the boats returned from the south. Nathaniel Ransom got back to the *John Wells* and told the mate, "I found the *Lagoda* [bark *Lagoda* of New Bedford, Captain Swift] and put our traps aboard of her with lots of other boats." Aboard the *Wells*, all hands had their dunnage packed ready for a start for the ships in clear water.

The second mate of the *Eugenia* came back, reporting heavy ice "all the way down to Icy Cape and two barks to the South of it, and blowing so hard, I could not communicate with them." He had left one of the mates and provisions landed on Icy Cape waiting for the wind to moderate until the latter could get off to one of the whaleships. As the boats returned, many of them were reloaded with food and sent south again. It was getting on in the morning. The log of the *Seneca* records: "The ice at present is crowding in on the land. The ship as she swings to her anchor, the rudder touches the ice. Barometer falling."

Now, one by one, the American flags appeared—the ships were setting their colors, the agreed-upon signal of abandonment, and they whipped smartly in the brisk southwesterly that had brought about the whole tragic business. Against the bleakness of the scene, the red, white and blue, conveying simultaneously to them all everything that was stable and ongoing even while it signaled the abrupt end of this massive misadventure, must have been a hard sight to bear.

The mate of the *Eugenia,* which was "bound solid in the ice," commenced "packing my boats and provisions over the ice a half mile to open water" and soon thereafter "left the good *Eugenia . . .* with her ensign flying at the mizzen peak."

The last shipboard entry of William Earle, first officer of the *Morgan,* relates that, ". . . all hope of saving ship or property is gone. If we save our lives, we ought to be satisfied and that should satisfy the world. To winter here might be possible, but not under present circumstances.

"We have neither the clothes nor the provisions, so to remain would entail an amount of suffering from cold and hunger and loss of life as would not justify anyone in attempting it. At twelve noon, paid out all of chain on both anchors and at 1:30 P.M., with sad hearts, ordered all the men into the boats and with a last look over the decks, abandoned the ship to the mercy of the elements. And so ends this day, the writer having done his duty and believes every man to have done the same."

And now, there were almost two hundred whaleboats in the narrow land water between Wainwright Inlet and Icy Cape, a unique funeral procession, if you will. For if there is nothing more alive than a wooden ship afloat, there is nothing deader than one without water. And it was without water that they last saw their ships, incredibly poised upon an ocean, yet with no ocean, silent and motionless hulls that darkened with the growing distance, the ensigns fluttering, union-down, the decks and davits empty, symbols of gracefulness suddenly arrested and grotesque, half swallowed in the angular jumble of the white world.

And so goodbye, and every dip of the oars, every ring of water from the dripping blades, every thump of the ice chunks along the planking of the boat made the incredible farewell more bitterly credible. Moreover, as difficult as it was to accept the abrupt end of what was—the termination of the whole cycle of everyday familiars, large and small—concern for the hours ahead immediately began to deny grief even its briefest hour, which was probably a good thing for them. The coming of Arctic afternoon, the raw southwest wind off the wintry sea reminded them with every passing moment that, in putting more than a thousand people in small boats upon this hostile sea in mid-September, they traveled a narrow threshold between safety and disaster.

There were no practical answers to these eventualities for the simple reason that they were already in a position so vulnerable to weather change as to be running a high risk. What they had done constituted a last resort and those of the thirty-two who had made the decision for the other 1,187 who believed in God—and most did—must have done the best they could to pray silently in their boats, being careful to display no sign of uncertainty or fear.

When they last saw the fleet they abandoned that day, the most northerly whaler, the *Roman*, lay in 70 degrees, 30 minutes north and 159 degrees, 30 minutes west. The most southerly vessel was at 70 degrees, 25 minutes north, and 160 degrees west. The fleet stood in a northeasterly line, slightly curved, and was anchored between Point Belcher, north, and Point Marsh, south. The vessels were, in some cases, five abreast, but usually, no more than three; they were at anchor in a strait, having to the west the heavy icepack and at the east the shore. Between the fleet and the shore was a narrow shoal and between this shoal and the shore was a stream of navigable water.

The vessels of the beleaguered fleet and their value, as published by the *Republican Standard* of New Bedford were: From New Bedford, barks *Awashonks*, $58,000; *Concordia*, $75,000; *Contest*, $40,000; *Elizabeth Swift*, $60,000; *Emily Morgan*, $60,000; *Eugenia*, $56,000; *Fanny*, $58,000; *Gay Head*, $40,000; *George*, $40,000; *Henry Taber*, $52,000; *John Wells*, $40,000; *Massachusetts*, $46,000; *Minerva*, $50,000; *Navy*, $48,000; *Oliver Crocker*, $48,000; *Seneca*,

$70,000; *William Rotch,* $43,000; ships *George Howland,* $43,000; *Reindeer,* $40,000; *Roman,* $60,000; *Thomas Dickason,* $50,000.

From New London, bark *J. D. Thompson,* $45,000, and ship *Monticello,* $45,000; from San Francisco, barks *Carlotta,* $52,000; *Florida,* $51,000; and *Victoria,* $30,000. From Edgartown, ships *Champion,* $40,000; and *Mary,* $57,000. From Honolulu, Sandwich Islands barks *Paia* and *Kohola,* $20,000 each; *Comet,* $20,000 and *Julian,* $40,000.

The fleet, including its catch of 13,665 barrels of whale oil, 965 barrels of sperm oil, and 100,000 pounds of bone, represented a loss of about $1.5 million.

Three of the abandoned ships—*Concordia, George Howland,* and the *Thomas Dickason*—were owned by George, Jr., and Matthew Howland. They constituted almost a third of the Howland whaling fleet, were worth a total of $168,000, and there was not a penny's worth of insurance on any of the three.

# 17.

# Rescue

Tell them all I will wait for them as long as
I have an anchor left or a spar to carry a sail.
—CAPTAIN JAMES DOWDEN, bark *Progress*

To the south, the seven ships on which the hopes and lives of the fleet of refugees depended had been in serious difficulties themselves for almost two weeks.

The Honolulu bark *Arctic*, the British bark *Chance* from Australia, and the American-registered whalers *Daniel Webster*, *Lagoda*, *Europa*, *Midas* and *Progress* were just north of Icy Cape on the first of September, when the ice closed, hemming them all in so that they had to anchor near the beach, with just enough water in which to float.

For ten days, until the eleventh of September, they were locked in hard, with no prospect of escape. But on the eleventh, the ice slacked sufficiently to allow them to work through it to a position about ten miles south of the Cape. The *Chance*, Captain Thomas H. Norton, was stove coming through the broken pack ice and was leaking badly.

Had the ice slackened even twenty-four hours earlier, allowing them to work offshore, they might have been gone beyond reaching, for it was on the eleventh that Captain A. N. Tripp of the *Arctic* "fell in" with Captain Fraser from the icebound fleet to the north, and learned of their disaster, and the seven masters, still working to get their own ships free, decided on their response to the appeal for help.

Economically, of course, the situation would be a major setback if they elected to effect the rescue. The seven vessels had taken only 3,070 barrels of whale oil and 27,981 pounds of whalebone; in other words, they had aboard $100,000 worth of product and hoped to make it about $300,000 if the whales remained plentiful for the rest of the season, and the ice allowed them to go whaling. There was another temptation. In 1870, the fleet's total catch had been 57,285 barrels of whale oil and 756,550 pounds of bone; this year, whatever the seven took would be all that was taken and the relatively smaller catch, no matter how well they did, was certain to drive up the price dramatically.

A congressional committee report, filed in a prolonged effort over many years to get federal remuneration for losses sustained by the rescue fleet because of its truncated season, offers an insight into the seven masters' reaction and response to the appeal for aid from the north:

These vessels [the seven] had arrived on the whaling ground fully prepared to prosecute the business of whaling. The whales were plentiful in all directions. Suddenly, while the prospects are so favorable for these vessels, a call is made upon them for succor by 1,200 shipwrecked seamen. These men have no money to promise for the rescue. They are shut up in the Arctic seas with an Arctic winter before them and the sure result, if they are not rescued by these whaling vessels is a slow and lingering death, either by starvation or cold.

The masters of these vessels have this alternative—on the one hand, of turning their backs on these men and leaving them to die, while they go on in the prosecution of their voyages and in making money for themselves, or, on the other, the sacrifice of their gains to rescue these shipwrecked sailors.

The choice was made without a moment's hesitation. The masters, with the full consent of all the crews, decided at once to abandon their voyages and to rescue these men, entirely regardless of self and without a murmur. They decided instantly to give up all hope of profit and all hope of reimbursing themselves for their expenses and to convey these men to a place of safety.

If they had adopted any other course, a cry of indignation would have gone up from the civilized world, which would have justly accused these claimants of a worse crime than murder, that of abandoning these men to a slow and horrible death.

But having made the decision to rescue was no more than half the task of bringing it about.

On the thirteenth, the ice came jamming in upon the seven once more. They escaped being imprisoned by the narrowest of margins, broke out of the pack through a combination of hours of hard labor and luck, especially handicapped by the fact that Captain Norton of the *Chance* considered his badly squeezed vessel barely seaworthy. Captain Thomas Mellen of the *Europa* wrote, "[We] got into clear water at dark by cutting the chain and letting the anchor go with forty-five fathoms of chain attached and then heaving her through the ice with cutting falls."

The rescue fleet was free of the ice, but now the weather turned poorer. There was a strong breeze from the north, a heavy sea building, and persistent rain and fog. Advance whaleboats from the abandoned fleet came off to them with provisions, unloaded their freight under the sloppiest weather conditions every time the wind let up a little, and then started back up the coast for more. They had as yet had no word on the day and hour of abandonment.

Meanwhile, the refugee fleet was inching its way down the coast. Of the first day, the fourteenth, William Earle, the *Morgan*'s first officer, wrote:

The wind being light at SE, were enabled to carry our sails, but one of the boats being very slow, our progress was not very rapid. All of the ships we passed were abandoned or their crews leaving in their boats. Hundreds of boats were ahead of us, as far as the eye could search. The last vessel we passed was the brig *Victoria* of San Francisco, hard aground and being well over on her side.

When about ten miles from our ship, met our third officer on his return. I turned him back, giving him part of our load and a part of the loads of the other two boats. As night came on, the wind increased and as darkness closed around us, heavy black clouds seemed to rest over us and it was not possible to see more than a few feet and we were in constant danger of coming in collision with the many fragments of ice floating in the narrow passage between the land and the main pack.

At 10:30 P.M., landed by a fire on the shore, where several boats were hauled up and made some coffee. While we were on shore, the wind began to increase, with some rain. Double-reefed our sail and single-reefed the jibb [sic], shoved off into the darkness at 11:30. The navigation was difficult and dangerous; we kept the land well aboard and sounding continually.

Some, including parties with the fleet's three women and five children, did not travel after dark, but camped on the beach behind some sandhills, rounded up a scanty supply of driftwood, and built a fire. The sailors dragged boats up into such lee as the low hills offered, turned them upside down and covered them with sails, to provide as much protection as possible against the frost and chill. Most of them, however, did nothing so elaborate but either kept on traveling or squatted by the fire, waiting for daylight.

The *Reindeer*'s boats, part way back to the ship after having put provisions aboard the rescue fleet, were hauled up on the beach when night came. E. M. Frazier, boatsteerer, said, "We made a tent of boat sails, built a fire to keep our feet warm, and went to sleep on the sand. About midnight, Captain Loveland [B. F. Loveland of the *Reindeer*] came into the tent, telling us we must retrace our course seaward, the whole fleet having been abandoned. The captain had all the remainder of his crew there in one boat, with two suits of clothes for each man."

To the south, the wind continued to blow hard; the *Midas* and *Progress* parted chain cable and lost anchors—the growing question was whether they could hold on long enough, for most of the whale-boats would not be arriving until the next day, and it was probable that they could not get everybody aboard before the sixteenth, at the earliest.

The second day out was the stiffest test for most of them. According to Captain Preble:

. . . the boats reached Blossom Shoals and there spied the rescue vessels lying five miles out from shore and behind a tongue of ice that stretched like a great peninsula ten miles farther down the coast and around the point of which the weary crews were obliged to pull before they could get aboard.

The weather here was very bad, the wind blowing fresh from the southwest, causing a sea that threatened the little craft with annihilation. Still, the hazardous journey had to be performed and there was no time to be lost in setting about it.

All submitted to this new danger with becoming cheerfulness and the little boats started on their almost hopeless voyage, even the women and children smothering their apprehensions as best they could. On the

BLO

SHOALS.     PLACE OF RENDEZVOUS.     ICY CAPE.

PLATE IV. "Abandonment of the Whalers in the Arctic Ocean, Sept. 1871." *Courtesy of The Whaling Museum, New Bedford, Mass.*

voyage along the inside of the icy point of the peninsula, everything went moderately well, but on rounding it, they encountered the full force of a tremendous southwesterly gale and a sea that would have made the stoutest ship tremble.

In this fearful sea, the whaleboats were tossed about like pieces of cork. They shipped quantities of water from every wave which struck them, requiring the utmost diligence of all hands at bailing to keep them afloat. Everybody's clothing was thoroughly saturated with the freezing brine, while all the bread and flour in the boats was completely spoiled.

The strength of the gale was such that the ship *Arctic*, after getting her portion of the refugees on board, parted her chain cable and lost her port anchor, but brought up again with her starboard anchor, which held until the little fleet was ready to sail.

## For First Officer Earle:

The fresh breeze [of the fifteenth] lasted till about 1:30 A.M., with a darkness almost black. Just as the wind began to die away, one of our boats came in contact with a small piece of ice, staving a hole in her bows. She hauled up on the beach and the hole being fortunately above water, soon repaired and followed.

Sent up rockets and assembled the boats at 3 A.M. and proceeded on our way. There were now four of us and at 7 A.M., hauled up to the beach, landed and made coffee and took breakfast, with what appetite may be imagined. At 8, embarked again. We were now about twenty miles NE from Icy Cape. About three miles to SW of where we landed, met another one of our boats, she being in search of us. We gave her part of our load and ordered her to take part of the loads of the other boats. Our company had now increased to five.

At 10:30, arrived at Icy Cape where, among twenty-five or thirty boats, found two more of ours, making seven in all. As the wind had again increased and would be nearly ahead after rounding the Cape, double-reefed our sails, ordered all of our boats to weigh and proceed.

Continuing along the land to SW of the Cape, which we did by beating, tacking and with oars between the ice and land till six or seven miles to South of the Cape, when we took dinner at 3 P.M. on a long fragment of ice. After some search, found an opening in the ice and with a fair wind, delivered the whole crew of the *Emily Morgan* on board of the ship *Europa* of Edgartown safely.

Nathaniel Ransom of the *John Wells* wrote, "September 15—Off Icy Cape at present. Strong breeze from SW. I've just [been] aboard of ship *Europa*, Captain Mellen, after being out in a hale and rain-storm, pulling and sailing, for last twenty-four hours. I had to throw my bomb gun, a box of bomb lances, with a musket and lots of ammunition, with several other things overboard and all cotes of Esquimaux garments."

Late in the afternoon, having taken aboard more than 153 ship-wrecked sailors, the *Lagoda*, Captain Stephen Swift, beset by the strong northerly that drove heavy rain before it, parted the chain of her port anchor and had to run for it. He went to the southeast until early the next day, when he was able to work back inshore with moderating weather, and take aboard more of the refugees.

As fast as the whaleboats were emptied, their occupants scram-bling up the ships' sides to get undercover and somewhat dried out. The boats were set adrift; there was no place for them on the already crowded ships. Of itself, this was a tragedy second only to the first abandonment; these scores of graceful white boats, built with pride, and even with love, whirl, spin and bob to leeward, afloat and empty for the first and last time, hastening down the wind to disaster, destined within minutes to become splinters on a barren beach. It was nothing for a boatman to watch.

To the *Europa* were assigned 280 refugees; to the *Arctic*, 250; to the *Progress*, 221; to the *Lagoda*, 195; to the *Daniel Webster*, 113; to the *Midas*, 100, and to the *Chance*, 60.

"There were not accommodations for more than forty men on board any of these ships," Captain William Kelley observed, "yet, in addition to their own crews, they had to divide up the 1,200 of us. Only seven of the splendid fleet of forty vessels that had left Honolulu less than a year before was left, but not a life had been lost."

On the sixteenth, the last boats' crews had been taken aboard and the rescue fleet weighed anchor. The last of the seven, the leaking *Chance*, was reported safely in Honolulu on November 22, 1871.

EUROPA.                                    DANIEL WEBSTE

SHIPS RECEIVING THE CAPTAINS O

PROOF.

MIDAS.          CHANCE.     ARCTIC.     PROGRESS.     LAGODA.
S AND CREWS OF ABANDONED SHIPS.

PLATE V. "Abandonment of the Whalers in the Arctic Ocean, Sept. 1871." *Courtesy of The Whaling Museum, New Bedford, Mass.*

# 18.

# Plunder

I have to express it as my decided opinion
that, unless such relief [as was provided by the
seven whale-ships] had been offered, fully nine-
tenths of the shipwrecked seamen must have
perished during the approaching winter.

—A. W. GREELY, head of the scientific expedition at Fort Conger, Discovery Harbor,
Lady Franklin Bay, 1881–1884, one of seven survivors of a twenty-five man team struck by
starvation and death when a relief party failed
to reach them.

There is a minor question as to whether the record of these events
should show that 1,219 people were rescued from the fleet or
whether it should be 1,218. Or perhaps it was 1,220 who abandoned
ship, and 1,219 sailed to Honolulu. The figure itself is not as important as the fact of the discrepancy, because what it means is that
when everybody else left, one white man stayed behind.

It is a pity that his name is not known. One suspects this is because his contemporaries did not take him seriously—a whaling
master of the abandoned fleet later remarked that the man was "not
quite right in the head." (But that has been said without justification
of a lot of people simply because they chose not to take the safer
way.)

Whatever he was, or whatever he thought, there was one among
those fleeing on the night of the fourteenth, one among those shivering on his haunches before the skimpy beach fire, who found the
hope of becoming rich stronger than that of being rescued. In the
darkness of that first night away from the ships, he deserted—if that
is the proper word for it—and made his lonely way back to the fleet,
spurred by the dream that he would have for himself some of the

million and a half dollars it represented. Presumably, he walked back, along the shore, and one has to admire his courage or marvel at his foolhardiness, depending on how you look at it.

You might say the wonder is that he did not have company. Surely others among the thousand must have toyed with the thought of staking a claim—all those ships, all that bone and oil, just for the taking. There was, after all, reason to believe that if one picked the vessel in the best circumstances, it might be possible to get through the winter aboard her. Everybody knew there was plenty to eat, especially just for a man or two. Tons of provisions had to be left behind.

During the winter, when the weather was decent, there would be chances to gather bone from the ships and stow it somewhere. When spring came, the ice went out, and the whalemen came up for the new season, one could lay claim to the whole business in the name of salvage, get a free trip home when they went south, and never have to work another day. Surely there would not be such an opportunity again.

But the fact is, this fellow did not have company. Whoever he was, he walked back into at least a fifty-fifty chance of death, as any of the survivors of the bark *Japan* could have told him.

To the Eskimo, the abandonment of the whaling fleet was a miracle of several parts. For him, it offered promise beyond measuring, as well as relief from his more efficient white competitor. Yet because he was a good man in the best sense of the word, his jubilation over his own prospects was characteristically tempered by compassion for the whalemen.

Why he should have had any is hard for a non-Eskimo to understand. The simplest explanation is that the compassion arose, not because it was logical, but because it was Eskimo—that is, the most natural thing for him to feel. The natives' reaction at Plover Bay when they were informed of the disaster was typical. They said to the white men: "Bad. Very bad for you. Good. Good for us. More walrus now," according to a letter signed "Whaleman" published in the New Bedford *Republican Standard*.

Yet as the whaleboat fleet made its way down the coast to safety on the fourteenth and fifteenth, the Eskimos—although obviously eager to get to the abandoned ships—did everything they could to help the fleeing whalemen and voiced their sympathy as well, which was remarked upon later by many of the refugees. Some of the natives spent hours helping the whaleboats get clear of the ice in the land water.

If they believed, as they might well have, that their gods had triumphed over those of the white man who had pushed them to the edge of starvation, they did not say it, nor did their actions reflect it. Perhaps they felt, watching those hundreds of suddenly humbled strangers, observing for the first time the white man in a poorer condition than they, that he had been punished enough.

Before the sixteenth, there remained a possibility that the whalemen might have to return to their ships for want of an alternative. The Eskimo knew this because they had told him so. But word of the deserted fleet had spread as fast as the whaleboats' journey down the coast, and the natives coming up from the south brought the news that the rescue had been effected. The white man was gone; he had left his ships behind, and the natural fear that had thus far restrained these perpetually impoverished people restrained them no longer.

It would be impossible to overemphasize the importance of this opportunity to the native. The abandoned fleet, insofar as it was a symbol of a civilization thousands of years more advanced in a material sense than theirs, represented to them the equivalent of innumerable lifetimes of effort, trading, inventing, manufacturing, hunting—in short, an unprecedented opportunity to remove themselves for the first time from the tight cycle of perpetual insecurity.

The fleet's cutting tools, cordage, canvas; its wood, whaling implements, and sharpening stones; its spikes and nails, wood-boring and shaping instruments; its firearms and fire-makers; its total treasure house of things they did not have, could not make, and perhaps had not even dreamed of, constituted a fantastic opportunity to make their daily tasks easier and even to gain instant affluence, in terms of trade. There was some reticence among them, especially among the older people, as to whether the white man's tools and weapons were as good as their traditional equipment, but generally, they respected very much what his harpoon, axe, and rifle would do. Besides, what they were getting was free, and to a society of traders, this was historic. Small wonder they believed that, of all people, they were the chosen people, because this unheard-of thing, the deliverance to them of a whole fleet of ships to do with as they would, had happened to them, and only to them.

By the eighteenth of September, the Eskimos had gathered along the shore between Wainwright Inlet and Point Belcher in some numbers; they must by then have concluded that the white men had been gone long enough so that it could be assumed they were not coming back. More important, the land water had frozen hard enough so that they could walk off to the icebound fleet. They swarmed aboard the ships, eager to get at the business of wrecking.

Over a period of several days, they carried away most of what was readily movable—spars, planks, boards, sails and rigging, even ornamental work from the after cabins. They carted away stacks of bone, got out a lot of the casked oil from belowdecks, and sledged and hauled off everything they could handle that could be used for fuel or hut building.

The plundering was done hastily, wastefully, and haphazardly, in part because they did not know how much time they had before worse weather would destroy the ships and everything in them. There also was a fever pitch about the operation; it was a profoundly emotional experience for them and, ultimately, there was a religious aspect which arose from an unlikely source, the ships' medicine chests.

The whaleship's medicine chest traditionally contained a relatively small quantity of liquor. When about to abandon their ships, the masters—either because of personal experience or hearsay—surely had in mind the point made by the anonymous author of the letters in the *Whalemen's Shipping List and Merchants' Transcript*. He had written:

*At Sea, December 22, 1852*

The last season, I was near the land only a few times and, in consequence, saw but little of the natives. They came on board off St. Lawrence Bay and at the Diomedes Islands. At both places, their chief desire seemed to be for rum, which they demanded by numerous signs and gestures. Notwithstanding all their gesticulations and grimaces, they got nothing but cold water from the butt.

Their great telegraphic feat was to feign to be exceedingly "set up"— that being a sign which they acted admirably—their excellent mimicry showing that they had had some experience. I cannot say how long it is since rum has been introduced to this people, but probably the Russians used it in barter for furs. I was not visited in the season of 1851 by any natives from the west coast of the straits, but I believe those on the east coast did not ask for rum. I hope our whalemen will do nothing to foster this appetite. These people are already wretched enough.

A schooner was there in 1851 from Hong Kong, trading for Walrus teeth and furs, with a plenty rum on board, which they gave in trade to the natives. This caused a good deal of trouble on board the whalers. A painful difficulty occurred between the natives and the crew of the ship *Armata*, in Bering Straits at the time of her loss, which ended in the death of eight natives and one English sailor. Several boats' crews were on board from other ships, one or two from English whalers, with a good many natives. Rum was the cause of the trouble.

I do not know upon whom the blame must rest, but I do earnestly entreat my fellow whalemen by every consideration of morality and of self-interest not to furnish these poor people with intoxicating drinks. They are very much in the power of the natives of these regions in case of shipwreck and it is only prudent to keep them as simple in their habits as possible.

Thus, when the masters left their ships, assuming that the natives eventually would come aboard, they destroyed by mutual agreement all the liquor. But because, when they left, there was a chance that they would have to come back and attempt to get through the winter there, they did not destroy the other contents of the ships' medicine chests, although they passed the word to the natives to leave these things alone.

Just as the language barrier prevented the ship captains from understanding the Eskimos' early warning that this year's ice would be different and dangerous, perhaps the natives did not comprehend.

Or maybe they did understand and the fact that the medicine chests were among the first things they went for simply reaffirms that humanity is what it is. Presumably some of them thought that anything in a bottle was drinking liquor, for the white man—the trader, if not the whaleman—had given them that idea.

So some of them drank the medicine, eagerly and indiscriminately, and some of them became violently ill. It is not recorded that any of them died, although it is a wonder if none did and perhaps there were, in fact, deaths, for there was not much of anybody, including the lone white man, to keep records. The illnesses they suffered were interpreted as the work of Toonook, the evil spirit, who, therefore, had to be driven out.

As a result, every ship on which an Eskimo was made ill by medicine was set afire. The Eskimos burned to the water's edge the *Florida*, aground on the Sea Horse Islands, and the *Gay Head*. The beautiful *Concordia*—presumably because of something as insignificant as a vial of ipecac—was reduced to a mile-high thrust of oily smoke struggling against the bleak sky, racing orange flame in the Arctic wind devouring Henry Purrington's "noble spread eagle," and finally, nothing but black ash and charred bones, stark in the interminable white wasteland.

Two weeks after the fleet was abandoned, a heavy northeast gale drove all but the ground ice away; that did not move. There will never be agreement on whether the vessels could have been gotten out during this brief interlude if they had not been abandoned, and as the years go by and the weight of evidence on each side yellows, becomes brittle, and falls into the mellow litter of eternity, it becomes increasingly unwise and unnecessary to attempt a judgment.

In any event, soon thereafter, a second gale stormed out of the north and rendered any alternatives academic. The only witness to what that was like, the lone white man, was, as with most whalemen, not poetic. He said simply, "Of all the butting and smashing I ever saw, the worst was among those ships, driving into each other. Some were ground to pieces, and what the ice spared, the natives soon destroyed."

For some period, this man, this lonely Macbeth, if you will, this irrational pursuer of an increasingly unachievable triumph, lived aboard the bark *Massachusetts*, which was shoved, slammed, and pushed by the ice easterly around Point Barrow, far to the north of anything that the fleet of 1871, when manned, had been able to achieve. During some period, as the weather permitted, he accumulated whalebone from the disintegrating wrecks.

That is, all but one were disintegrating wrecks. The *Minerva*, grounded at the entrance to Wainwright Inlet, was miraculously spared as all the others were miraculously destroyed; she remained as sound, as tight, as shipshape, as when abandoned. All the others, even those not burned, were in bad shape. The *Thomas Dickason* lay on her beam ends on the bank, bilged and full of water. The *Seneca* was dragged up the coast, her rudder carried away, bowsprit gone, and bulwarks stove; the *Champion*, aboard which the thirty-two masters had drawn up those last fateful documents, was ashore and broken to pieces; the *Emily Morgan*, with her masts gone, was ashore, and so were the *Reindeer* and *Kohola*. The rest were either carried away by the ice or crushed.

Most of the male Eskimos eventually turned against the white man who wintered with the wreckage of the fleet. Why, he apparently did not say; perhaps he chose not to. One suspects it came down to the division of a fundamental commodity—the whalebone; he accumulated a great deal of it and at some point, the native men took most of it away from him and threatened to kill him, although which came first is not clear. He was quoted as saying, "They set out to kill me, but the women saved me and afterward, the old chief took care of me. A hundred and fifty thousand dollars would not tempt me to try another winter in the Arctic."

In the spring of 1872, "pretty well used up" and possessing nothing but the clothes in which he stood, he walked five days across the ice and was rescued by returning whalemen. He told them that the Eskimos, contrary to their living pattern of centuries, had fled inland, fearing that the white men would punish them because they had burned some of the vessels.

# V

# The Last Meeting

# 19.

# "Deviation from natural law"

> In reading the account of the vessels being
> wrecked by the ice in the Arctic, it occurred
> to me that it might be worthwhile to state a
> fact which may not be generally known. Last
> winter in that region was the most remarkable
> for the long continuance of the cold and its un-
> usual severity for the past several years . . .
> Under ordinary circumstances, this is not likely
> to occur again.
> —CHARLES BRYANT, special agent, U.S.
> Treasury Department, November 20, 1871.

Considering that George and Matthew Howland wrote off
$300,000 forever as a result of the disaster of 1871 (or as much as
$1,500,000 in today's money) and that, as far as they were con-
cerned, one could not speak of whaling with the same respect again,
one might wonder why they did not get out of the whale fishery,
then and there. They were, after all, otherwise so shrewd, so boldly
cautious.

There were some good reasons for staying, not the least being,
as Mr. Bryant commented, the belief that ". . . this is not likely to
occur again." After all, everybody knew that Arctic whaling was
a gamble and despite the terrible setback of '71, the Howlands had
to admit that they had been largely winners for five decades. Besides,
who is ever willing to admit that, in terms of the inevitable odds, the
time has come to pay?

The thinking of the Howland brothers is undoubtedly reflected
in the remarks of their peers as reported in interviews obtained in
New Bedford by a reporter of the Boston *Post* and published in
November, 1871:

I called early this morning at the office of Messrs. Swift and Allen, one of the largest houses in the city who date the commencement of the firm name as far back as 1842.

The countinghouse has an air of neatness and thorough business arrangement. On a shelf over the door were a dozen metallic boxes containing the private papers of the ships of which the firm are agents, with the names painted in large letters on the front.

Conspicuous among the rest were the words *Fanny, Massachusetts, Eugenia* and *Elizabeth Swift,* all of which were lost in the sweeping down of the ice from the polar sea.

Mr. J. Swift, Jr., was seated in his private office on the left of the entrance. He is a gentleman of probably fifty-two or three winters, black hair and side whiskers tinged with grey, very courteous and cordial in manner, thoroughly posted in the whaling business and has met with general good success. Previous to the recent disaster, the firm had lost only one whaler in the ice, the *Monongahela,* Captain Seabury, about fifteen years ago. They had one burned by the *Alabama* during the war and they indulge the hope that they will be reimbursed for that loss from the treasury of John Bull.

I found Mr. Swift reading a clipping from one of the Boston evening papers and about commencing an article in reply to a statement concerning the danger of whalers remaining too late in the season in the high latitudes, and which conveyed the impression that the merchants of New Bedford were carrying on a business that was dangerous to both life and property.

The following conversation ensued:

*Q.* Mr. Swift, I am most desirous to learn your views upon the recent disaster to the whaling fleet, its probable effect upon that business, and upon the business of New Bedford generally.

*A.* It is a hard blow for us to bear and though there is no cause for any alarm concerning the future, some people think we have been doing a business that was altogether unsafe and that the risks were more hazardous than any insurance company should take. Now the fact is that the records of our insurance companies show that the risk on merchantmen bound for the Indian Ocean are as great as those on whalers bound for the Arctic regions. The currents and unknown shoals of the Indian Ocean destroy quite as many ships as the ice of the Arctic. The fleet did not stay too long up there at all, for they were a full month earlier than usual and were making their way up instead of back. Risks on whalers were really the best that could be got.

*Q.* How about the losses? They will be paid by the insurance companies?

*A.* Certainly, although they will all have to be borne right here in New Bedford. Our companies have all been mutual for several years and the stock has been subscribed for by our own people who own ships or parts of them, so that the owners are, in reality, insurers, too. So long as the companies attended strictly to their legitimate business in insuring whalers and such merchantmen as went from our own port, they have been safe and at all times before they adopted the mutual plan, they declared twenty and sometimes as high as forty percent. This was the experience of the old Pacific Company, until it got to taking out-of-town risks, which caused it to wind up in 1866. Had they not gone out of town, they might have kept ahead until now and divided 33⅓ percent annually. As it was, the merchant risk used up all their assets and forty percent on their stock. Our companies do not take merchant risks now except such as are connected with our own trade. We expect to hear about once a year that one or two ships have been crowded in the ice or have been lost, but that's a very small proportion of the whole fleet.

*Q.* What will the effect be upon the whaling business?

*A.* Well, sir, the truth is, there have been more in the business than it would support, so there will be no increase. This is the fourth time since 1842 that the business has been flat, when we could make nothing, but we have built it up every time. Lately, the market had been crowded and the price kept down. If Provincetown had never gone into the sperm oil trade, we would have sold for enough to make a handsome profit.

*Q.* The ships in port will be fitted out for next season?

*A.* Undoubtedly, and many will fit out who had no thought of doing so before. Ships will be more valuable, prices of oil higher, and, having had a bad year, there is every reason to expect that the next will be a profitable one. Another thing in favor of fitting out now is that all the officers and captains of these ships are coming home. There will be two captains to every ship and all anxious to go. We have just received a dispatch from the officers of our ships saying they want to go out again just as soon as possible. If there was any such danger as some people like to think, it is not likely these men would want to go back again. I think the trade will be better in a few years and we shall make up the loss.

The reporter from the *Post* also interviewed Captains Daniel Wood and S. C. Smith, and he described them as:

. . . old whalers, having been in the Arctic regions and cruised among the icefields ten or twelve winters in succession and, I believe, without ever sustaining more than trifling damage to the ships in their charge.

They said that the ice begins to break up generally in July and leaves the water open in different places so that the whalers can cruise along without danger. In a few weeks, the ice masses in one grand barrier, so to speak, to the northerly, stretching from opposite Point Barrow far out into the polar sea. The ships can make their way back and forth with no more risk than one undergoes in traveling by sea between New Bedford and Boston.

Point Belcher and Wainwright Inlet, between which the fleet was hemmed in, are not so far north by ninety miles as the whalers usually go and there must have been a strong northwest gale to start the ice barrier down toward shore so much earlier in the season than generally happens. In 1868 and 1869, the ice opened about this time of the year, but that seldom occurs.

The fleet were evidently trying to make Point Barrow and were perhaps in such a hurry to reach the whaling grounds they forgot to keep a sharp lookout; the gale came up suddenly and shook them up before they could get out.

The whalemen understand the lay of the ice just as well as they do that of the land. When it comes down, it moves very slowly, not more than a mile an hour, and at this season, the winds are very light, though often it is moved faster by the force of the current. Whalemen do not generally leave Point Barrow till the first of October; many stop later and do not depart until the latter part of October; they seldom venture beyond Point Barrow for fear of being crowded inshore by the ice, from which there is no escape when once it begins to close in.

Captains Wood and Smith both thought the disaster would deter no whalemen from going on another voyage; the disaster was merely one of those deviations from natural laws against which all precautions are futile. Such an event would not probably occur again in a lifetime.

In a majority of cases where whalers were jammed in the ice, it was caused by negligence or recklessness of danger on the part of the masters. The ambition to make a successful catch often induced men to brave hardships on sea as well as on land and the proportions of mishaps on the latter were far more numerous than on the former.

Thus, it was quite clear that what had happened was an unfortunate combination of "recklessness," "deviation from natural law," that it was not likely to occur again and that, in fact, by abruptly reducing the number of ships engaged in whaling, it actually placed the owners of the ships that remained in a much better competitive position. Quite clear, indeed, except that just one thing went wrong.

After the losses of '71, the Howland fleet stood at seven vessels: *Rousseau, Onward, Java, Desdemona, Clara Bell, George and Susan,* and *St. George.* Only five years later, in the fall of 1876, the "deviation from natural law" occurred again, trapping in the ice thirteen of that season's Arctic whaling fleet of twenty vessels. Of the thirteen ships abandoned, four—*Onward, Java, Clara Bell,* and *St. George,* representing $126,000, not including bone and oil aboard that probably totaled a third again as much—belonged to George and Matthew.

Call it then, something in excess of a half-million nineteenth-century dollars that the brothers wrote off in a five-year period, and that doesn't even count the potential earning power of those lost ships.

The golden moment that began with *Concordia,* that faltered with her burning, had ended.

# 20.

# Down to Meeting

Thee must remember that I am in a crippled condition and cannot do as I could when I had seven or eight ships and oil and bone at good prices. All I expect or hope to do is, by great economy, to struggle through . . . I am obliged to curtail and deny myself in order to be faithful and honest . . . We have just had an application from New York to purchase the *Rousseau,* but I don't imagine it will result in a sale.
—Letter of Matthew Howland, October, 1879

By 1880, Matthew Howland was resisting advice to outfit *Desdemona* for another whaling voyage. It was his view, born of tightening financial circumstances, and falling behind $5,000 to $6,000 a year, that "it will not do for me to send her out on any uncertainty. I cannot afford to lose $10,000 or $20,000 more. It seems to me, until we are more enlightened than we are at present that it will not be prudent, to say the least, to fit out *Desdemona.*"

His wife, Rachel, in letters to her children written in November, 1882, revealed their deteriorating situation. "I should have written thee sooner, but have been very busy taking care of Father, who has been quite ill. Second day night, he suffered extremely; in the morning, as soon as I could, I got the doctor here and he staid three or four hours trying to allay the pain, which he finally accomplished by several doses of morphine injected into the arm.

"W. is still trying to get this place, but says he will not give over $30,000, while Father asks $40,000. What shall we do? I think Father is very poorly indeed and very low-spirited. What shall we do with him or for him?

"We have been greatly exercised about selling our house. C. has not treated us handsomely. So he thinks he could just gobble us up and turn us out of house and home at his pleasure. And so the big fish eat up the little ones.

"C. said that, of course, he must make very considerable changes in the house, that though it might do very well for 'your wife' etc., it would not suit Mrs. C. at all as it is etc.

"It was quite edifying to see the air with which Mrs. C. surveyed the rooms. 'I wonder, Mrs. Howland, you do not occupy the southeast room for yourself,' etc. And when I took her into the north parlour, she remarked that it would do nicely, with more light, for her son's studio. 'You know, he is something of an artist' etc.

"So it was quite an ordeal for me and I felt rather overdone by the tour of inspection of the premises. I stated several times that the house was not on the market, was not on exhibition, but she insisted on continuing her investigations into every room and closet, while I, acting as Cicerone, was considerably rasped.

"F. is in New York now. If thee should meet him and he should speak of the matter, I do hope thee will not encourage him to buy this place. I do believe, if we are patient, we can get along without the help of stranger-capitalists. They are so sure we are at their mercy that it makes me angry.

"Father has gone down to Meeting and I am alone . . ."


"Meeting," of course, as with all else, was not the same. Those who attended were, as always, outstanding members of the community, yet there were fewer of them and the impact of the Friends upon New Bedford, its dreams and destinies, was much diminished. As if the Meeting were inextricably linked to whaling which, in a certain sense it was, it rose and fell with the industry. There was much more to it than that, of course,—schisms tore the flock, new restlessness arose in the young, the ethnic base of New Bedford's population broadened, the haste of manufacturing was superimposed upon the leisureliness of the maritime community. Yet the result was that the golden era of meetinghouse and countinghouse coincided. And when the golden time was gone for one, it was gone for the other as well.

It was not only a matter of smaller Meeting attendance; for George and Matthew, still faithful, it was increasingly a matter of remembering what had been and was no more, of recalling who had sat in the pews now empty. Certainly this is true of all old men. Yet for them it was more completely true and with more reason. For they had lived long enough—or events had proceeded with sufficient rapidity to a most unpredictable and unlikely end—so that they were forced to witness the winding down of everything, including some things that had taken generations to build and given the false promise of being everlasting.

Matthew died in 1884 at the age of seventy. And Rachel, intelligent Rachel, still proud and indomitable, looked the other way while lawyers swarmed over her property. It says something that Matthew, even when he had nothing left to do but die, signed one of his last letters "yours in haste." Haste was thrifty and haste was a habit.

Late in the year in which Matthew Howland died, Matthew Barney of the Society of Friends wrote to George Howland from Nantucket, reporting on the effort to revitalize the Meeting there that had begun coincident with the launching of *Concordia*. Of the situation on the island, to which George had devoted much energy and attention, Barney observed, "I cannot see that any other conclusion could be come to in regard to our meetings that they must cease for the present . . .

"I felt much regret that I could not report that a few of us would continue them yet so I cannot report.

"The meetings have been very acceptable to many of our summer visitors, making same acquainted with a friends' meeting for the first time and showing them that we were neither Shakers nor of the Spiritualist class, but believed in the old-time doctrines of the New Testament . . . Although I cannot tell any positive results, yet I believe that much quiet sympathy and comfort remains behind."

Four years later, George's wife of sixty-one years, whom he had married in the Friends' meetinghouse, also died, this event reminding that three children, all sons, had been born of this union, of whom two died in infancy, and the third at the age of twenty-eight, in 1861.

The total number of members of the New Bedford Monthly Meeting had dropped to slightly more than three hundred. Edward T. Tucker, chairman, writing in behalf of the Sandwich quarterly meetings, advised, "Dear Friends: Our monthly meetings are directed to substitute the word 'may' for the word 'should' in Line 15 on Page 100 of the Book of Discipline."

George lived on, yet stood increasingly still. While he became more and more preoccupied with remembering, the New Bedford that he had known and shaped became something else; the industry on which it had relied to accumulate unprecedented wealth and worldwide stature was dismantled, or perhaps unraveled is a better word. The people in the streets—once, he would have known many of them and their genealogies as well—became strangers. It was a total process of tumbling in and tumbling down; not as with Dorian Gray, for George was, by constrast, a good man, but the diminution of the microcosm was quite as inexorable, quite as thorough.

What reasons? Probably more than anything else a refusal to take note of the speed at which time ran, and at which change was taking place—a stubborn determination to believe that what had been regarded as sound and good, what had worked for a long time, would forever be sound, good and workable. Surely George was not alone in adhering to such views as these; his errors of judgment and consequent defeat beneath the burden of his compounding woes are understandable in human terms.

For George was, of all things, human. This was the man who, obviously reaching out to share the moment's richness, wrote from Civita Vecchia, ". . . I exchanged bunches of flowers and put out lights with people whom, of course, I never saw before and probably shall never see again." So it is not unreasonable to suggest that as he sat in the meetinghouse in those last years—a structure that was venerable, dignified, functional and unadorned, as was he—his reflections were both philosophical and poetic.

It might well have occurred to him that perhaps no Howland

would ever again occupy the mayor's office in New Bedford's City Hall (and none has). Surely, he suspected that members of the Society of Friends would no more exert the principal influence in shaping the political, social, and economic destinies of the city. (They have not.)

The restless, increasingly prosaic community, each day losing some of those qualities with which George had felt at home, was ironically born in considerable measure of his own good works, his shrewd leadership, his ability to influence people because they liked and respected him. Thus did such major shapers as the gasworks, railroad, expanded water system, the new textile mill come to be. Yet what all of these brought as social by-products made the city, his city, no less foreign to him.

The broadening of the popular base, with new tongues and new beliefs, had brought an inevitable end to the theocracy that had produced *Concordia*. It became increasingly obvious, for example, that it was simply not enough to assume one's philanthropic obligation to the poor, to the unchosen many; there were pressures abroad asserting that there ought to be fewer poor and more chosen. One sensed the emergence of new voices in the land, although it was still difficult to know, especially for an old and tiring man, what they were saying or to what gods, upper or lower case, they paid obeisance.

The wind of change was not only raw, its breath shook the foundations. Obviously, the rich did not necessarily remain rich. George might have grimaced at that thought; to do him justice, what diminution of fortune had meant most to him was frustration at not being able to finance those good works to which he had devoted much of his life. And if the rich did not necessarily remain rich, perhaps the poor did not necessarily remain poor any longer, either, so it was more difficult than ever to know who was Chosen and who not—even Friends differed now as to how one told the difference or whether, in fact, one could or should.

One might argue that those who sent the whaleships to sea measured the oceans in dollars and cents, rather than in fathoms, and they did. Nevertheless, they plumbed the universe in their fashion;

they had daily intercourse with tides, winds, continents, the change of season, far-off peoples, and other civilizations. And even though they may have been dogmatic in their approach to things and inflexible or worse in their approach to people, there can be no questioning the depth of the religious beliefs by which the best of these men lived and through which, for better and for worse, they justified their lives.

Finally, George became ill, although that is not precisely the word, either, because there was no acute disease, but rather a general weakening of the system. In the early winter of 1892, marking the last phase of his being, he lapsed into occasional delirium and at such times spoke nothing but French, which he had learned in boyhood and for which he retained a lasting affection. Psychologists or psychiatrists may make more of this phenomenon, but one suspects it was principally the mind's return for comfort to those years when golden moments were commonplace.

George died on February 18, 1892, and his funeral services, attended by a large number, were held in the Spring Street Friends meetinghouse, where he had been a bulwark from the time it was erected, nearly seventy years before. Matthew led, George followed.

George's biographer, William L. R. Gifford, wrote, "His kindliness, his willingness to aid, and his conspicuous integrity had combined to render him one to whom positions of trust came almost in the nature of things. A retiring disposition and a strong love for home surroundings made him shrink from, rather than seek public office.

"Yet he believed it his duty not to withhold his services when his fellow-citizens desired to make use of them; and to the demands of any position to which he was called, he gave an even more conscientious attention than he bestowed upon his own private interests.

"Although a man of quick decision and of positive convictions, he never refused to give ear to opinions which differed from his own; and however resolute and determined his subsequent action, it was not undertaken without a careful consideration of every point at issue. The duties of each recurring day he faithfully performed, and the result was a life of singular usefulness and merit."

George's long and deservedly effulgent obituary omitted only one important detail. As long as four years before his death, he had had to sell the home in which he had lived for more than a half-century and he had died virtually penniless.

The *Rousseau,* whose figurehead old George had chopped off and thrown into the mud, outlasted them all, even the *Desdemona.* After more than fifty years of highly successful whaling for the Howlands (she took nearly four thousand barrels of oil between 1866 and 1878), when the tide of fortune and industry turned against her, she lay idle at the foot of New Bedford's North Street, near where the Howland wharf had been.

Even *Rousseau* was no longer Howland owned. On February 17, 1882, Matthew had written, "Yesterday we closed with Aiken and Swift for *Rousseau* and *Desdemona* at $8,300. They don't wish anybody to know it until they go to work on them in the spring. It is a *very low* price, but we did not think we could keep them longer. We commenced several weeks since, asking $10,000, they offering $6,000 for both of them. *Finally,* we came to the result as above . . ."

But *Rousseau* did not go to sea. The New Bedford *Standard* reported that the Girard interests in Philadelphia, from whom she had been purchased by the Howlands, long years since, wanted to buy her back. Stephen Girard, merchant, banker, and philanthropist, was dead, but he had left funds for establishment of a college bearing his name, for "poor white orphan boys" and its officers thought *Rousseau* would make a schoolship, to be operated in conjunction. "But nothing ever came of the idea," the *Standard* reported.

She lasted, did the *Rousseau,* having been built, as they said in those days, "on honor"; when her bottom was replanked in 1879 (at which time she was seventy-nine years old), her floor timbers were "as firmly on her keel as when constructed." Lying there unused, she became a landmark which is, after all, something that should never happen to a ship. Stump-masted, her topmasts and topgallant masts having long since been sent down for the last time; weatherworn and frayed, the tag ends of such rigging as remained dangling

*Rousseau* (inboard) and *Desdemona* (outboard) in their final years. Note *Rousseau*'s billethead. *Courtesy of The Whaling Museum, New Bedford, Mass.*

untended; paintless and scoured by storms, she was picked clean, over the years, of everything that was valuable and movable.

In her last days, she was called the oldest ship afloat; tramps lodged in her after cabin in their ordure. Finally, she was broken up for the metal in her and her bones were towed to the north end of Pope's Island in New Bedford Harbor, where they were laid to rest. Even then, not exactly to rest, because for some time, the ingenious visited the wreck and removed some of her timber, making walking sticks of it and frames for pictures of the *Rousseau* in her better days, both products being readily salable.

In the end, wind, weather, mud, wash of the tide, landfill, time, and the inevitable surrender of substance eliminated even her bones, good old bones though they were.

# Acknowledgments

A respectable amount of the material in this book, accumulated during more than thirty years in and about New Bedford, Massachusetts, from people, many of whom are no longer living, derives from the oral, rather than the printed word.

I have relied heavily upon the conversations of a lifetime with my father, Joseph Chase Allen of Martha's Vineyard (happily, still sound of limb) who is an authority on whaling, and the author of biographies of Captains George Fred Tilton and Hartson H. Bodfish, Arctic whalemen, both of whom were family friends of long standing.

I have also depended upon interviews with Captains Ellsworth West of Martha's Vineyard and Charles Chace of Westport, outstanding ship handlers and whalemen; with Amos Smalley of Gay Head, who inherited the Indian's traditional skill with the harpoon, and who "ironed" a white whale; with John McCullough, who, with his father, sent a number of whaleships to sea from New Bedford, and with numerous mates and foremast hands from Southeastern Massachusetts, especially among those of Portuguese descent (excellent sailors, by heritage and instinct), who went whaling for one or more voyages.

I am indebted as well to many talks with William H. Tripp, former curator of the Old Dartmouth Historical Society and Whaling Museum in New Bedford, who went a-whaling one trip and wrote about it; with the late Reginald B. Hegarty, curator of the New Bedford Free Public Library's Melville Whaling Room, who went whaling with his father, Captain William C. Hegarty, and whose conversation and writings have been most helpful in this project. For a valuable introduction to the Eskimo and the Arctic,

I am obligated to Rear Admiral Donald B. MacMillan of Province-town (last survivor of the Peary North Pole expedition, author of an Eskimo-English dictionary and several volumes on the Arctic, and founder of a school for Eskimos at Nain, Labrador), whose biography I was privileged to write.

I am further indebted to Richard C. Kugler, director of the Whaling Museum in New Bedford, and to Philip F. Purrington, its senior curator, for their continuing interest and generous assistance; to Rita E. Steele, librarian of the Millicent Library in Fairhaven, especially for material on the Society of Friends; and to the Rhode Island Historical Society, Providence, for permitting me to read the Monthly Meeting records of nineteenth-century Friends in New Bedford.

For general information on the period, I read issues of the New Bedford *Republican Standard* for the eighteen months beginning January 1, 1871, including articles reprinted there from 1871 and 1872 issues of the Boston *Advertiser*, Boston *Journal*, Boston *Post*, Alta *Californian*, Hawaiian *Gazette*, Hawaiian *Friend*, and the Honolulu *Gazette*. My untiring wife, Phyllis, transcribed thousands of words of tape-recorded notes from these and other sources.

# References

## THE LAUNCHING

The New Bedford *Mercury* files of 1867 provided the bare bones of what *Concordia* was like Mr. Hegarty's classic *Birth of a Whaleship* (which few today could write) revealed how the pieces of the ship went together, and a letter from my father was the source of information on wood selection, tools employed and shipyard worker implements, practices, and attitudes.

## BOOK I

CHAPTER 1: For the writing of all chapters dealing with the Eskimo and the Arctic, I found it especially helpful to read Robert F. Spencer, "The North American Eskimo," Smithsonian Bulletin 171 (Washington: 1959); Herbert L. Aldrich, *Arctic Alaska and Siberia* (New York: 1889); William H. Dall, *Alaska and Its Resources* (Boston: 1870), and P. H. Ray, Lieutenant, U.S. Army, "Report of the International Polar Expedition to Point Barrow, Alaska" (Government Printing Office, Washington: 1885).

The bowhead's deep-diving ability when frightened is as stated in the volume produced by Tre Tryckare, *The Whale* (New York: 1968).

CHAPTER 2: Also helpful in these chapters on the Arctic and its people was *Cruise of the U.S. Revenue Cutter* Bear *and Overland Expedition* (Washington: 1899), especially the report of Lieutenant D. H. Jarvis, commander of the overland expedition.

CHAPTER 3: In addition to recollections of Captains Bodfish and Tilton and to my father's biographies of them, I have relied on some log excerpts concerning Arctic whaling from vessels commanded by Captain William C. Hegarty, the father of Reginald.

## BOOK II

CHAPTER 4: For both this and the next chapter, I benefited much from the reading of Leonard B. Ellis, *History of New Bedford* (New Bedford: 1892); Frank Walcott Hutt, *History of Bristol County* (New York: 1924), and Daniel Ricketson, *History of New Bedford* (New Bedford: 1858), the latter especially being a charming and scholarly presentation.

CHAPTER 5: Of singular value here and in Chapter 3 of this book were the writings of Llewellyn Howland III, unpublished manuscript, Boston,

1964; Allan Clapp and R. H. Thomas, *History of the Friends in America* (Philadelphia: 1920, reprint edition); William L. R. Gifford, *George Howland, Jr.* (New Bedford: 1892); David Moment, "The Business of Whaling in America in the 1850s," a research report, Graduate School of Business Administration, Harvard University, 1957; New Bedford Monthly Meeting records of the Society of Friends, and the booklet by Zephaniah W. Pease, "Centenary of the Merchants National Bank," New Bedford, 1925.

CHAPTER 6: The *Republican Standard* reprinted Edward King's Boston *Journal* article; the parenthetical references are from 1871 issues of the *Standard*. Waterfront items are from the *Standard* and from Alexander Starbuck, *History of the American Whale Fishery* (Waltham: 1878).

CHAPTER 7: The eulogy for Matthew Howland by Daniel Ricketson appeared in the December 1884 issue of *The Old Colonist*, then published monthly in New Bedford; excerpts from letters by Matthew are from the unpublished manuscript of Llewellyn Howland III, and from Matthew's letter book, 1858–1879, Baker Library, Harvard University. For further information on owner-agent countinghouse operations, I am indebted to a symposium held at the Whaling Museum, New Bedford, on November 11, 1972, and especially to an address, "J. and W. R. Wing, a Case Study of the Management of a Whaling Firm," by Martin J. Butler, professor of history, Southeastern Massachusetts University, and to related commentary by Mr. Kugler and Mr. Purrington.

CHAPTER 8: In this and other chapters, I have relied on the *Whalemen's Shipping List and Merchants' Transcript*, published in New Bedford (filed in the Melville Whaling Room), and particularly on issues appearing in 1852, 1871, 1872, and 1874.

## BOOK III

CHAPTER 9: John G. Abbott's journal, written aboard the *Huntress*, is in the possession of the Whaling Museum and was first called to my attention many years ago by Mr. Tripp, then curator of the museum.

CHAPTER 10: I found helpful a reading of Olive Wyndette, *Islands of Destiny* (Rutland, Vermont: 1968) and James Jackson Jarves, *History of the Sandwich Islands* (New York: 1843). John F. Willoughby's article, "Hawaii, the Pearl of the Pacific," was distributed by the American Press Association in 1898; the copy I found in the files of the New Bedford *Standard-Times* had been filed, but never published in that newspaper.

CHAPTER 11: Letters by Captain Frederick A. Barker, concerning the loss of the *Japan* and his winter with the Eskimos, were published in both the *Republican Standard* and the *Whalemen's Shipping List and Merchants' Transcript*, in 1872.

## BOOK IV

In the chapters relating to the Arctic whaling season of 1871 and events leading to and concerning abandonment of the fleet, I have depended on logs or published log excerpts, found in the Whaling Museum and the Melville Whaling Room, of the following ships: *Eugenia, Emily Morgan, Contest, Henry Taber, Seneca,* and *Fanny.*

The two documents signed by the whaleship masters were reproduced in several publications, including Starbuck's *History of the American Whale Fishery,* and the *Republican Standard;* Captain Henry Pease's account of the storm in the season of 1870 was republished by Starbuck (probably from the *Whalemen's Shipping List*). Charles Bryant's comment on the Arctic weather was printed in the *Republican Standard* in 1872, and Nathaniel C. Ransom's journal, written aboard the *John Wells,* is in the possession of the Whaling Museum.

The log of William Earle, first mate of the *Emily Morgan* (customarily, the first mate, not the master, kept the ship's log), including its supplement, covering the whaleboat trip from Wainwright Inlet to Icy Cape, was of special worth; one must conclude from his writings that he was an able and conscientious officer. The value of the abandoned whaleships is that published by the *Republican Standard,* apparently based on estimates furnished to the newspaper by owners, agents, and insurance companies.

I also found helpful the account of the Arctic disaster published in *Harper's Weekly,* New York, December, 1871, and reprinted in Starbuck. Additional events related to the rescue I obtained from Whaling Museum scrapbook clippings, *Republican Standard* reprints from Hawaiian newspapers, excerpts from the log of the *Lagoda,* and Committee on Claims Report, U.S. 46th Congress, 2d Session, Washington, April 8, 1880.

CHAPTER 18: Captain William H. Kelley, who commanded the *Gay Head,* revisited the area where the fleet was abandoned the following year and wrote a letter home concerning what he saw and was told. This was reprinted in part in Starbuck, in the *Republican Standard,* and in the *Whalemen's Shipping List* and accounts for knowledge of the one man who was left behind in 1871. Kelley's account is substantiated by a letter published in the *Whalemen's Shipping List* of October 1, 1872, from "Smithers, first officer of bark *Live Oak,*" who also reported on the condition of the wrecked fleet, the burning of some abandoned ships by the Eskimos, and who noted that "the boatsteerer that stopped here [for the winter of '71] is still alive . . ."

## BOOK V

Letter excerpts in the final chapters are from the unpublished manuscript of Llewellyn Howland III; some of the references to George Howland, Jr., derive from Mr. Gifford's biography of him, and facts relating to the *Rousseau* in her last days are from a story written in 1905 for the New Bedford *Standard* by William G. Kirschbaum.

# Index